The ~~KEHL~~ Inconvenient Truth

It's Warming:
But it's *Not*
Human-Caused CO$_2$

M. J. Sangster PhD

This book is dedicated to my friend
Terry and brother Jim.

Acknowledgements

My deep appreciation for the help I received from Lorne Dey, Darren Hansen, Ted Payne, Duane Gentner, Al Miller, and Howard Durant, Bethany Payne (Ted's cousin), Bob Bidwell, and Elsbeth Walkley.

Darren provided excellent feedback in the early stages of the book and Ted later in the project, and Duane was kind enough to edit several chapters. Al and Howard provided feedback in the early stages of the project and were subjected to weekly monologues as my research evolved. Bethany and Bob helped me with final touch-ups, and Elsbeth saved the day when I was having such a difficult time finalizing the book layout. Elsbeth also provided graphical support.

A special thanks to Lorne who produced many of the illustrations, designed the cover, and published the book. Without his support, I would have been left holding an unpublished manuscript.

I am grateful to my wife, Sheila, who put up with 2 years of my rambles as I plowed through research papers talking to her about my findings and her responding with more enthusiasm than I know she had.

And finally, to my daughter Tess who is always an avid supporter of me.

Table of Contents

Preface

In autumn of 2016, I was sitting with a friend in Liverpool, England, my birthplace, a friend who I know to be very intelligent, when he said to me, without provocation, *"I believe in global warming, y'know Mike."* For some reason, I believe he thought I was a skeptic. Of course, at the time I didn't know enough about the subject to take a position one way or the other, a point I made. But I was curious as to how he arrived at his position.

When I asked him why he believed in global warming, and what did that mean, he gave what I have since found to be the normative retort, *"it's pretty obvious;"* after all, 2015 and 2016 were very warm summers. It's obvious that the Earth is nearer to the Sun in summer than it is in winter – except, it isn't! Science doesn't deal in *the obvious.* I told him that I would research the subject and get back with a more scientific basis. I expected my reply would be a short summary of my findings, but the gravity of the subject matter was so compelling that it took me about two years of research to prepare a response – this book.

Early in the process I became a skeptic, not initially of the science, for I knew nothing of that for several months, but of the Alinsky-style tactics of global warming activists, and the constant beat of the drum from: politicians demanding that the *science is settled*; politicians demanding that there is *consensus* among scientists that the science is settled and, probably the most motivating reason was; claims by politicians and media that *those disagreeing with them are evil, and that laws should be enacted to silence them.* These had a strong odor of intimidation, propaganda, and even a form of fast fascism; Orwellian, in fact. Add to those, the numerous *holocaust* references associated with skeptics, and I concluded *politicians and media doth protest too much*; viola, another skeptic was created - me!

And it wasn't long into my research that I formed a skeptical position on the hypothesis that human-caused (anthropogenic) carbon dioxide (CO_2) was the cause, or even a major contributor to global warming. I found no dispute that the Earth has experienced an overall warming trend beginning at some point in the 19th century, around 1850-ish and continuing through today. And I found no dispute that during that time CO_2 levels have increased and that humans have contributed to the increase, and there is little dispute that human-caused CO_2 is a *factor* in climate change. I found no skeptic scientist who advocated maintaining the status quo; they agree that we should be researching energy alternatives. Nor did I find any skeptic scientist who claimed to *know* the cause of climate change. So why is there a debate? Because it's not about science, it's about power and control; the power to redistribute wealth by controlling energy policy. This is not my conclusion, it is a stated objective by key anthropogenic global warming (AGW) advocates, including a former chair of the Intergovernmental Panel for Climate Change (IPCC).

The warmest year in recent history was, arguably, 1934 when CO_2 levels were significantly lower than today. There is a debate whether 1998 was warmer, but it doesn't matter since the difference would be statistically insignificant, and so even if 1934 was the second warmest year, the argument is sound, CO_2 levels were about 30% less than they are today. And, according to the World Meteorological Organization (WMO), the hottest recorded temperature *ever* was on June 10, 1913 when Death Valley, CA, registered 134°F (56.7°C), when CO_2 was almost at pre-industrial levels.

In the 1970s it was so cold that we were warned of an *Impending Ice Age*. And it is well documented that from about 1998 through 2014 there was no increase in global warming; a period that even staunch AGW advocates acknowledge was a *pause* in global warming. During these periods, CO_2 steadily increased. It was during the period of global temperature stability that global warming became *climate change*, and human CO_2 emissions became responsible for warming, cooling, no change, and every natural climate event, hurricanes, tornadoes, earthquakes, tsunamis, etc.

AGW activists have convinced the public that there is a *normal* climate and that prior to the Industrial Revolution the Earth was in its natural state, when in fact the Earth's climate has, over its more than four billion years, always changed, and numerous times much more extremely than has occurred over the past century or so. **There has never been, nor will there ever be, a *normal* climate.**

Climate changed in the past, it is changing now, and it will change in the future. It changed before humans existed and it will change after we're gone. Earth was a frozen mass of ice for more than 50 million years, and the relatively recent warming phase followed several hundred years of cold, frigid temperatures, a period known as the Little Ice Age (LIA), which extended from the 15th to the 19th century. Prior to the LIA, the Medieval Warming Period (MWP) occurred, from the 10th to the 14th century during which Vikings settled in Greenland, only to leave or perish during the LIA. And before the MWP, during the Roman Empire days from about 250 BC to 400 AD, there was a warming period when temperatures were as high as, or higher than, at any time during the 20th or 21st centuries. According to Robinson and Robinson (2012, *Global Warming; Alarmists, Skeptics and Deniers – A Geoscientist looks at the Science of Climate Change*), from a geologist point of view the recent warming is not at all unprecedented and that over an extended period of time is a *"rather minor fluctuation."*

A key component of the debate that can be and is often (mis)-used to support a position is the time frame over which data are displayed. For example, when the intent is to show a rapidly increasing temperature trend, the author of a paper selects a suitable period that shows this to be the case; from 1980 to 1997 for example, when there *was* significant warming. But if the time period was from 1945 to 1975 it would show a significant cooling, an *Impending Ice Age* in fact.

Another commonly used tactic to mislead, is presenting correlation between two variables as a *cause-and-effect* relationship. Correlation statistics are useful for providing valuable decision-making information, such as whether product sales correlate with increased advertising, but can be misused when presented as cause and effect. And correlation statistics can easily be misunderstood; for example, a strong correlation between two variables of say 80% leaves 36% unexplained, not 20% as might be inferred.

There are books written on "How to Lie with Statistics" (Huff, 1954) and how to "Statisticulate" (Rowe, 2013, et al), showing how to misinform and/or manipulate people by the use of statistical material. To quote W.W. Norton (1954; ISBN 0-393-31072-8),

"The secret language of statistics… is employed to sensationalize, inflate, confuse, and oversimplify…. without writers who use the words with honesty and understanding and readers who know what they mean; the result can only be semantic nonsense." I leave it to the reader to conclude how politicians use statistics.

In researching this complex topic, I read 26 books, hundreds of blogs, articles, and journals, and watched scores of YouTube presentations, and I watched Al Gore's movie, *An Inconvenient Truth: A Global Warning*, and read his book *An Inconvenient Truth: The Crisis of Global Warming*. My education comprises a B.Sc. and M.Sc. in Electrical Engineering, an MBA, and a PhD in Human and Organizational Development and, prior to retiring, I was a

Fellow of the Institution of Electrical Engineers for more than 30 years. My PhD dissertation was *Judgment and Decision Making under Conditions of Uncertainty* in which I examined a number of psychological biases and their effects on rational, normative decision-making processes. I have taught university undergraduate and master degree students in Decision Making, Statistics, and Business Research. I designed and developed a simulation language (*Interactive Simulation Interpretive System*) for studying dynamic systems described by their differential equations, and I have conducted numerous trade studies as part of the system engineering approach to facilitate decision making of complex simulations. I have the qualifications to understand scientific research papers but,

I believe that it's impossible for a layperson, including me, to advocate with *strong conviction* a position on the science of this subject; it is simply too complex and multi-disciplinary, and the science community is too divided. That does not, however, excuse our blindly accepting any single organization or person's position; we owe to it ourselves to research the subject and be informed, and not leave it to politicians or media as our only, or primary, source of information. While I cannot take a strong position on the cause of climate change, I am convinced there is ample evidence clearly demonstrating that focusing only on human-caused CO_2 as the single or even main driver is not only problematic from a scientific viewpoint but, probably more egregiously, from a policy point of view.

When asked what I would like as a *take-away* from readers of this book, my response is that I would like for advocates of the AGW position to conclude that perhaps they should be less *certain* in their beliefs on climate change, and particularly that CO_2 is the cause of the warming trend over the past hundred years or so. Chapters 1 and 2 should create doubt and uncertainty about the AGW position which, if valid, would not require the behavior described in those chapters, and Chapter 4, section 4.5 includes numerous statements of uncertainty by IPCC scientists about their understanding of the physics, interactions, and feedback processes of clouds and aerosols, two significant contributors to climate. Chapter 5 shows that the models on which AGW claims are based are invalid.

The purpose of this book is however, not to defend the use of fossil fuel but to show that it is not proven to be the cause of global warming or climate change. Pollution associated with both the extraction and burning of fossil fuel is a major source of environmental and public health concern and, in my opinion, the sooner we can convert to a clean alternative energy source, the better.

A second *take-away* would be that skeptics not refer to the *entire* debate as a *hoax;* such dialogue is not helpful and creates unbridgeable trenches. I believe it is important to concede that the temperature trend *is* one of warming. This is not problematic since the data are recorded from the end of the Little Ice Age, so of course we are warming, and that is a good thing; Chapter 7 provides evidence that cold weather kills about 10 times more people than warm conditions. The debate should not be whether the Earth is warming, but whether it is caused by the human contribution to atmospheric CO_2 or mostly by natural events.

However, while I would like skeptics to be careful in their use of the term *hoax* and not apply it to the entire debate, I do agree that the war on fossil fuel and human emissions as the cause of global warming *is* a hoax, a malicious deception, a non-supportable argument dissembling the real objective of a *cause* which is more about destroying western civilization and redistributing wealth. This is clear from quotes by leaders of the AGW movement presented in Chapter 1.

The expectation that AGW advocates and/or committed skeptic *hoaxers* will abandon their positions of certitude is however challenging. According to neurologist Dr. David Rock (2009, *Psychology Today*), our brains don't like uncertainty; it's like a type of pain, something to be avoided. We crave certainty of information in the same way we crave other basic needs such as safety, food, etc. But I still can hope that some will abandon certainty and side with Voltaire who said: *"Uncertainty is an uncomfortable position. But certainty is an absurd one."*

Given then, that psychology is against achieving my primary take-away goals I hope the book provides more open-minded readers with sufficient information to engage in meaningful dialogue on climate change and global warming at a macro level.

The book is organized as follows: an Introduction, which puts into perspective the multi-disciplinary nature of climate-related science, and why claims of consensus are meaningless; Chapter 1 presents my views on how I believe politicians and media manipulate the public – while this section targets climate change, many of the same processes have been and are employed by all types of ideologues to achieve political and social agendas. Chapter 2 describes the Intergovernmental Panel for Climate Change, its charter, policies, its lack of scientific integrity, and the involvement by its members and supporters in scandals that include manipulating scientific results. It also presents examples of deception by AGW advocates, including government agencies. Chapter 3 provides a history of climate variation, showing that 20th and 21st century temperature and carbon dioxide levels are neither extreme nor unique.

Chapter 4 deals with the Science. It discusses naturally occurring events that affect climate: the Sun; tectonics; atmospheric and oceanic circulations; ocean oscillations including El Niño/La Niña; clouds and aerosols, and; greenhouse gases, with discussions on water vapor, carbon dioxide, climate sensitivity, and the carbon cycle - a continuous carbon exchange process that maintains the Earth's equilibrium. Chapter 4 concludes with a macro view of the Earth System.

Chapter 5 discusses the inadequacy of IPCC climate models, which are the sole basis for the catastrophic predictions of climate over the next hundred years yet, cannot replicate the recent past, i.e., they are not valid. Chapter 6, Economics, addresses supply and demand of fossil fuel, the enormous cost and futility of the effort of adopting non-fossil fuel energy policies, and the *Precautionary Principle*, a decision-making process that empowers the IPCC et al to make decisions without compelling scientific evidence or cost/benefit analyses; i.e., it allows them to ignore rational decision-making practices. While a discussion on economics has nothing to do with creating a case for doubt or uncertainty about the role of CO_2, I believe it is important to understand the cost, human and financial, of eliminating fossil fuel at this time. Chapter 7, Discussion, begins with a Q&A section followed by my thoughts on the AGW hierarchy and a few bullets on how we should proceed. Chapter 8 is a Conclusion.

Chapters 1, 2, 3, 5 and 6 end with chapter summaries. Because of the complexity of the material, Chapter 4 (Science) has summaries following each section.

The case for uncertainty about the role of human-caused CO_2 in driving climate is made in Chapters 1, 2, 3, 5, and 7. Chapter 4 adds to the case by showing the relative insignificance of CO_2 in the face of natural events, but could be skipped in its entirety, although I'd recommend reading sections 4.5 and 4.6. Section 4.5 discusses clouds and aerosols, their significance in affecting climate, and the uncertainties of climate scientists in understanding their behavior. Section 4.6, greenhouse gases, should be reviewed since that is the most often discussed aspect of the science. Other than that, Chapter 4 provides a foundation of climate knowledge that would allow the reader to engage in a reasonable debate and may inspire an interest in further research.

Since, as a layperson in this field of climate science, the only *knowledge* I have of the subject is from authoritative sources, this book contains many references and citations from experts which I hope doesn't make for difficult reading. Where there are quotations, I have taken the liberty of emphasizing various parts that I consider to be most important, and in some quotations, I have added parenthetical clarifications.

Introduction

This is clearly a very difficult topic to understand and more so to arrive at a reasoned position. When Newton's theory of gravity was challenged by Einstein's General Theory of Relativity in 1915, after more than 200 years as the prevailing theory, it was startling and emotional for *Newtonians*, and there were many skeptics, worldwide. Then, in 1919, Arthur Eddington demonstrated the validity of Einstein's theory by experimentation; he observed and documented the bending of light around the Sun during an eclipse, and that together with a *little* bit of math, settled it, QED. Once the evidence was demonstrated to support the theory it was accepted by the scientific community as the new theory of gravity - but still subject to scientific scrutiny.

There is no experiment that can confirm the cause and effect of climate change. And there is no way of isolating human-caused (anthropogenic) CO_2 from natural atmospheric conditions; **all claims that anthropogenic CO_2 is the primary cause of global warming are based on non-validated models.** Models that cannot replicate recent past climates are used to predict the climate decades into the future, and they are used to support non-viable energy policies in Europe, Australia, New Zealand, and to a lesser extent, the USA.

As discussed in Chapter 4, there are numerous natural cycles that influence climate; they can last from months and decades to hundreds and even thousands of years. Furthermore, there is no single climate "system" scientist *per se*; climate analyses and evaluations are conducted by a wide array of multi-disciplines. Dr. Barry Brooks compiled a list of the core scientific disciplines that have been primarily responsible for developing our current understanding of climate change and its implications. His list, presented in an August 2008 blog is as follows:

Atmospheric and Physical Sciences: climatology, meteorology, atmospheric dynamics, atmospheric physics, atmospheric chemistry, solar physics, historical climatology
Earth Sciences: geophysics, geochemistry, geology, soil science, oceanography, glaciology, paleo-climatology, paleo-environmental reconstruction
Biological Sciences: ecology, synthetic biology, biochemistry, global change biology, biogeography, eco-physiology, ecological genetics
Mathematics, Statistics and Computational analysis: applied mathematics, mathematical modeling, computer science, numerical modeling, Bayesian inference, mathematical statistics, time series analysis.

Plimer (2009, *Heaven and Earth*) adds several more disciplines, astronomy, geochronology, tectonics, archeology, palae-ontology, palae-ecology, and sedimentology.

And in 2013, a NASA article about the effects of solar variability on Earth's climate makes the statement,

"Understanding the Sun-climate connection requires a breadth of expertise in fields such as plasma physics, solar activity, atmospheric chemistry and fluid dynamics." It continues, *"No single researcher has the full range of knowledge required to solve the problem."* And this is for just one aspect of climate science.

Neither Brooks, Plimer, nor the NASA article include many sub-disciplines such as the humanities and social science, economics, or engineering, all of which contribute greatly to an understanding of the broader issues, especially with respect to the potential impacts of climate change and our ability to manage and mitigate them. But the above is sufficient to underscore an important point: scientific understanding of global warming and climate change impacts are not the domain of one field called *climate science*, and it is certainly beyond the understanding of any layperson including political advocates. It is in fact beyond the understanding of any one scientist, as NASA et al clearly acknowledge.

Scientists may be expert at one or two of the above areas but nobody has the credentials in all or even most of the subjects. Hoffman and Simmons (2009; *The Resilient Earth*) say,

*"All disciplines from the natural sciences are involved in climate study, to the point where **gaining a detail overall understanding of climate is impossible.**"*

This alone makes claims of *consensus* meaningless; consensus among whom, geologists, astrophysicists, biochemists? In fact, the nexus of climate science disciplines and resultant IPCC apocalyptic predictions are statisticians and modelers. Not climatologists, meteorologists, or any of those climate-related scientists mentioned by Bond, Plimer, or NASA, but those who decide what data to use and how to use them and those who create millions of lines of computer code to simulate the Earth and its environment.

The *theory* that global warming and/or climate change is caused by humans is not a theory in the scientific sense, it is at best a hypothesis postulated by its supporters. Scientific inquiry begins with a hypothesis that is then tested under controlled conditions and in accordance with the *Scientific Method.*

The scientific method is a process that has characterized natural science since the 17th century. It consists of an initial hypothesis, followed by experimentation to test the hypothesis under controllable and repeatable conditions. If results do not support the hypothesis then the hypothesis is discarded or modified to reflect the data. However, as mentioned in the Preface and discussed more fully in Chapters 3 and 5, there are numerous periods when the hypothesis that anthropogenic CO_2 is the cause of global warming/climate change is not supported by the data, yet the hypothesis remains dogmatically intact.

The Intergovernmental Panel on Climate Change (IPCC) was established in 1988 by two United Nations Organizations, the World Meteorological Organization and the United Nations Environment Program, with a charter,

"to assess the scientific, technical and socioeconomic information relevant for the understanding of the risk of human-induced climate change."

Note that it doesn't seek to determine *if* human-caused carbon dioxide is the primary cause of climate change it has, by virtue of its charter, already made that *a priori* conclusion. Karl Popper, generally regarded as one of the greatest philosophers of science, said,

"Whenever a theory appears to you as the only possible one, take this as a sign that you have neither understood the theory nor the problem which it is intended to solve."

And, as we shall see in Chapter 3, CO_2, natural and anthropogenic combined do not show a controlling cause-and-effect relationship with global temperature at any time throughout the history of the planet including the time since the Industrial Revolution. In fact, there is not a causal correlation between CO_2 and global average temperature over any meaningful statistical period, including the past hundred or so years. That is not to suggest that CO_2 and temperature are not related, they are, but the evidence strongly suggests that **natural events dominate climate variability**.

The IPCC does not conduct primary research itself but rather synthesizes the most recent developments in climate science every five to seven years, preparing periodic assessment reports (AR). The ARs are generally too technical for the general public and politicians to follow, so it generates a *Summary for Policymakers* (SPM) that claims to highlight the most critical developments in language accessible to the world's political leaders. However, as we shall see in Chapter 2, the SPM is used as a tool to advance *the cause* rather than to summarize the technical work, and in many instances fabricates scientific results.

If the study of climate change was conducted strictly as a scientific endeavor, while there would be differences among scientists, there almost certainly would not be a public debate. There would be advocates of certain theories and research would follow the scientific method. Peer reviews would be conducted and research findings challenged for one or many reasons; methodology, data sources, measurement techniques, statistical analysis, etc. But never would results be expressed as irrefutable facts; Never! **Science is never settled**. There is *established* science, $E=mc^2$ for instance, but not settled science. Established scientific theories are testable, falsifiable, repeatable, and observable, and they can explain past phenomenon and predict future phenomenon. The AGW position does not meet any of these conditions.

There is an irony associated with alarmist claims that the science is settled when the IPCC invokes the *Precautionary Principle* for its decision-making guidance to State members. This is discussed in Chapter 6, but by definition **the precautionary principle provides a policy framework for decision-makers in the *absence* of settled science**.

Furthermore, if the science is settled why are billions of dollars continued to be spent on confirming that the science is settled? This is of course, a rhetorical question. The answer is obvious; climate change has evolved into a mega-billion dollar industry with an inertia that will resist any attempt to diminish its business base. Of course, the *climate industry* has no interest in the science being settled; that is simply a political talking point.

Unfortunately, climate science has been politicized and, as with most political issues, the public is polarized, to a large extent based on political bias: in the USA, democrats generally agree that CO_2 is the problem; republicans that it's not. The psychological term for this behavior is *cultural cognition*, the theory that we shape our opinions to conform to the views of the groups with which we most strongly identify. There are many psychological terms used to describe individual and group behavior that the reader will recognize throughout the book, mostly in Chapters 1, 2, and 7. They include:

Maslow's Hierarchy of Needs – Psychologist Abraham Maslow defined a hierarchy of needs in which the most basic needs (air, water, and food) must be satisfied before being concerned with other needs. The next level, safety and security are needs that must be *mostly* satisfied before an individual considers other needs such as belonging, being part of a family or group. In the context of climate change, needs include maintaining a career and working in an environment of acceptance. To achieve this one must be part of a group, one that overtly supports the dogma, *the cause*.

Skinner's Operant Conditioning – Operant conditioning is a method of learning that occurs through rewards and punishments for acceptable or unacceptable behavior, respectively. Through operant conditioning, individuals make an association between a particular behavior and a consequence. According to the principle, behavior that is followed by pleasant consequences is likely to be repeated; supporting the AGW hypothesis for example is rewarded with research grants, group acceptance, celebrity, and advancement opportunities. Behavior followed by unpleasant consequences is less likely to be repeated (lack of grants, threats, criticism, shunning, etc.).

Sense of Control: There is an innate and apparently biologically motivated need to feel a *sense of control* over our domain; when we feel out of control, we experience powerful and uncomfortable tension. From an evolutionary standpoint, if we are in control of our environment, we have a far better chance of survival. If climate change is due to natural events, we have no control. Aware of this psychological phenomenon, demagogues instill a *sense of control* in the minds of the public by providing simple solutions to what they portray as a threat to survival – reduce or eliminate human-caused CO_2.

Groupthink – Psychologist Irving Janis described the concept of *groupthink*. Groupthink occurs when a group with a particular agenda makes irrational or problematic decisions because its members value harmony and coherence over accurate analysis and critical evaluation. It is a kind of thinking in which maintaining group cohesiveness and solidarity is more important than considering the facts in a realistic manner – it encourages *in-group* consensus, shielding the group from dissenting views, and holding negative attitudes towards the *out-group*, those with differing viewpoints. In the context of global warming/climate change, the out-group are skeptics, or more heinously, *deniers*.

Conformity - The most common and pervasive form of social influence; an extension of groupthink, it is the tendency to act or think like members of a group in attitudes, beliefs, and behaviors to group norms; be a team player and don't rock the boat. Chapter 2 is replete with examples of this behavior.

Group Polarization - The tendency for a group to make decisions that are more extreme than the initial inclination of its members –

it results in extreme risk-aversion or extreme risk-seeking, depending on group composition. Examples of this behavior can be seen in Chapter 2, including Climategate, where scientists went to extreme lengths to conceal and or alter scientific data that refuted their in-group ideology.

Tribalism – Incorporates characteristics from each of the above areas. For thousands of years, tribal cohesion was essential to human survival. To maintain cohesiveness tribes needed a common concept that could not be disproved, e.g., a god, or the AGW hypothesis! Almost all societies or tribes had a common concept of the supernatural, and almost all of them saw their worst threats, hunger, disease, natural disasters, or a loss in battle as a consequence of disobeying a god. Religion therefore fused with communal identity and purpose, and the idea of people within a tribe believing in different gods was incomprehensible; such heretics would be killed. Today, tribalism's most egregious manifestations are extreme nationalism – we are superior; religious dogmatic exceptional-ism – our god is real, we're right you're wrong, and; other isms such as groups based on race, etc. And of course, climate change. Demagogues have created tribes with the same passion and commitment as those engaged in the "isms." People psychologically identify with a group, the *in-group*, and at the expense of those not in the group, *out-groups*. Tribalism mostly perceives situations as zero-sum; there has to be a winner and a loser. In the context of climate change, laypeople are tribal in their beliefs; climate change to them is as unifying as religious beliefs. There are quotes in Chapter 2 of alarmists calling for the death penalty for *heretics*; i.e., members of the *out-group*, skeptics.

Who do we mostly associate with advocates of the AGW hypothesis: EU political leaders, UN political leaders, Al Gore, John Kerry, CNN, et al, not scientists? And because climate change is a political issue it has all the attendant acrimony and intolerance that divisive politics engenders.

Chapter 1
Political and Media Manipulation

Unlike most scientific research, climate change has become highly political. In fact, as mentioned in the Introduction, most climate-related positions are espoused not by scientists but by political figures; and politicians always have an agenda.

Climate has become the bedrock of western governments energy policies, and unfortunately, of scientific funding. In his farewell address, January 1961, President Eisenhower warned against science being controlled by governments via government funding. (The U.S. had funded a huge scientific effort for the Manhattan Project and President Truman's administration had continued to fund scientific research).

Clearly his concern was that to maintain the independence and integrity of scientific research it must not be subjected to governmental bias and/or agenda. Regrettably, that's precisely what is occurring. According to the *Science and Public Policy Institute*, from 1989 through 2014 the U.S. federal government funded scientific research of climate change to the tune of $32.5 billion.

Donald W. Miller (2007, *The government grant system: Inhibitor of truth and innovation. Journal of Information Ethics*) says that since the establishment of a Research Grants Office in 1946, the state has taken control of information; it uses federal tax money to fund and control research through the grant system and it forms mutually advantageous partnerships with industry and the academic community, which do its bidding - President Eisenhower's concerns have materialized.

Dr. Miller laments that the western tradition of information ethics dating from ancient Greece to the 20th century, characterized by freedom of speech and inquiry, has been co-opted by the government. He makes the point that knowledge advances by questioning, i.e., being skeptical of accepted paradigms, not conforming, saying,

"The state thwarts this by requiring its tax-funded scientists to conform to the official establishment view on such things as global warming...."

With massive funding to support its agenda, the government is able to exploit the fears of an uninformed public, for when it comes to climate change, the public *is* uninformed. And so long as the populace remains uninformed they are subject to government manipulation. Thomas Jefferson had concerns about government control of an uninformed people,

"Every government degenerates when trusted to the rulers of the people alone. The people themselves, therefore, are its only safe depositories. And to render even them safe, their minds must be improved to a certain degree." --Thomas Jefferson: Notes on Virginia Q.XIV, 1782. ME 2.207, and *"Though [the people] may acquiesce, they cannot approve what they do not understand."* --Thomas Jefferson: Opinion on Apportionment Bill, 1792. ME 3:21.

The following sections discuss how politicians and their media acolytes shape the minds of the public.

Knowledge: In his 2006 movie, *An Inconvenient Truth*, to draw attention to the dangers of thinking you know something that actually isn't true, Al Gore quotes Mark Twain: *"What gets us into trouble is not what we don't know. It's what we know for sure that just ain't so."* There's a delicious irony in this, since there is no evidence that Mark Twain ever said or wrote this quote. Yet, apparently Mr. Gore knows this for sure! However, the quote, based on a version by Josh Billings, an American humorist who, in 1874 had written: *"It is better not to know so much, than to know so many things that ain't so"* is insightful, sends a valuable message regarding our *knowledge*, and presents a nice segue to the epistemological premise we should consider when discussing important and highly complex matters such as climate change: **what do we know, and how do we know what we know?**

Epistemologists generally recognize at least four different sources of knowledge: (1) intuitive - based on feelings rather than hard facts, takes forms such as belief, faith, intuition, etc.; (2) authoritative - review of the professional literature, etc.; (3) logical - arrived at by reasoning from one point that is generally accepted to another point that is new knowledge, and; (4) empirical - based on demonstrable, objective data, which are acquired through observation and experimentation, such as Eddington's eclipse observation. While researchers often make use of all four of these ways of *knowing*, for the layperson trying to form a position on such a complex topic as climate change, authoritative knowledge (2) is the only reasonable primary source of knowledge, but hopefully augmented by logical reasoning (3).

Clearly, a layperson cannot apply logical reasoning to more esoteric subjects such as theoretical physics or the study of subatomic particles; in those cases, all our *knowledge* comes from authoritative subject matter experts and, in such cases, it doesn't matter if what we believe is true or not. But it does matter in the case of climate change and, while our knowledge is necessarily based on authoritative sources we can apply reason. And we should apply reason, not only to extend our views of data provided by authoritative figures but to question the basis of the source; **authority does not validate a scientific conclusion, only evidence can do that**.

Unfortunately, because of a person's ontology, or view of the world, they are likely to seek authoritative *knowledge* from those with whom they generally agree on most other issues (same tribe or *in-group*) such as fiscally conservative, socially conscious, religious, atheist, etc., and not apply logical reasoning. This of course is not knowledge, but belief.

Beliefs are reinforced by the *confirmation bias*, a cognitive bias where there is a strong tendency for people to treat data selectively and favor information that confirms their beliefs. According to a Gallup poll (2001), 20% of Americans believe in astrology, that the position of the stars can affect people's lives. Their beliefs are *confirmed* by reading supportive material (by other believers), attending meetings that are conducted by similarly minded people, latching on to the occasional coincidence, etc. Cognitive biases are tendencies to think in certain ways that can lead to systematic deviations from a standard of rationality or good judgment.

Another such bias, relevant to this discussion, is the *recency bias*. Recency bias is the phenomenon of a person most easily remembering something that has happened recently, rather than balancing data over a more statistically appropriate period. For instance, when *An Inconvenient Truth* was released in 2006 it was in the wake of the then recent horrific Tsunami of 2004 and the 2005 hurricanes Dennis, Katrina, Wilma, and Rita. We remembered the television coverage vividly and these events had a significant emotional effect on viewers, which almost certainly influenced how the movie was received and internalized by the public.

The general public *learns* from news media, politicians, and other generally self-serving activists who shape their beliefs. There is, in most cases, little if any critical thinking, and once a belief is internalized, there is almost nothing that can change a person's position. In 1894, in *The Kingdom of God is Within You*, Leo Tolstoy wrote,

"The most difficult subjects can be explained to the most slow-witted man if he has not formed any idea of them already; but the simplest thing cannot be made clear to the most intelligent man if he is firmly persuaded that he knows already, without a shadow of a doubt, what is laid before him."

Vocabulary: What is true in terms of gaining knowledge is also true for shaping our lexicon; we adopt epithets and banalities from public figures who are often, if not mostly, attempting to manipulate others to adopt a particular agenda, usually one that enhances their political power base or increases their wealth, or as is usually the case, both. Homophobe, sexist, xenophobe, racist, liberal, communist, fascist, hater, et al. Labels, connected by platitudes, seem to be the main source of *reason* in arguments among the general population when discussing topics in which they have no empirical or even properly sourced authoritative knowledge, climate change included. Labels are a distraction; they prevent rational discourse; their intent is to shut down opposing views; they are in fact a manifestation of Orwell's 1984 *Newspeak*.

And then there is *denier*, a term used for climate change skeptics, which is obviously intended to debase and vilify opposing views by employing a word associated with the holocaust, the nadir of human behavior. Not since a *consensus* shut down Galileo has such intimidation in the name of science prevailed. Name-calling is a well-known propaganda technique that links a person, or idea, to a negative symbol. Propagandists who use this technique hope that the audience will reject the person or the idea on the basis of the negative symbol (in this case the holocaust), instead of looking at the available evidence. Al Gore actually skipped the nuance of association when in 1992 in *Earth in the Balance: Healing the Global Environment*, and quoted in the *National Review*, actually said,

"Global warming is an environmental holocaust without precedent.... The evidence of an ecological Kristallnacht is as clear as the sound of glass shattering in Berlin." Wow!

Gore's obsession with associating climate skeptics with Nazi Germany was again obvious when, in 2007 at the UN Climate Change Conference, Bali, Indonesia, he said,

"One of the victims of the horrors of the Third Reich in Europe during World War II wrote a famous passage about the beginnings of the killings, and he said, "First they came for the Jews, and I was not a Jew, so I said nothing. Then, they came for the Gypsies, and I was not a Gypsy, so I said nothing," and he listed several other groups, and with each one he said nothing. Then, he said, they came for me."

The *victim* Gore was quoting was Martin Niemöller, a prominent Protestant pastor who emerged as an outspoken public foe of Adolf Hitler (after earlier being a supporter) and spent the last seven years of Nazi rule in concentration camps. Niemöller finished the statement with *"and there was no one left to speak for me."* His main point was not to be silent in the face of atrocities.

Apart from again trying to associate skeptics with the Third Reich, what on Earth was Gore talking about? Was he associating scientific skepticism, the cornerstone of scientific inquiry, with atrocities? It seems as if he carries around a number of quotations in his mind and slides them into his presentations for emotional effect, regardless of the context. They just sound good.

But in any case, skeptics are the ones under attack from public figures and media, and as disclosed later in this chapter, many attacks are so egregious that if made against politicians would result in a visit from the Secret Service; **AGW extremists have advocated murdering skeptics.** So in the context of the quote, that silence can be egregious, where are the voices condemning the verbal abuse and physical threats made against skeptics?

Gore's association of skeptics with undesirable motives continues: in an interview with USA Today (January, 2018), he is quoted as saying the following:

"The world is at the beginning of a sustainability revolution that has the magnitude of the Industrial Revolution. We saw it with the marriage equality movement. We saw it earlier with the civil rights movement. All these movements have lumped along very slowly with an agonizingly slow pace, and then all of a sudden there's an inflection, and people say, 'Oh, I get it.'"

His implications are clearly that (in addition to emulating the Third Reich) skeptics are against equality of marriage and, more heinously, against civil rights. He did the same thing in his movie when he said that we (the U.S.) had beaten slavery and desegregated schools; again associating skeptics with those who were in favor of slavery and segregation.

A statement by Joseph Nye Welch (1890 –1960), the American Lawyer (who served as the chief counsel for the U.S. Army, while it was under investigation for communist activities by Senator Joseph McCarthy's Senate Permanent Subcommittee on Investigations) comes to mind when he famously asked McCarthy, *"At long last, have you left no sense of decency?"* **Mr. Gore, at long last, have you left no sense of decency?** Surely if you have the science behind your claims you do not need to vilify those with opposing views. Possibly you, as a politician who gets elected based, in large part, on personal attacks on opposition candidates (first against your own party member(s) who challenge for candidacy, followed by whomever the opposition candidate may be) may not be aware, but **skepticism is integral to scientific enquiry.** He has obviously been prepped on the vital role of skeptics in science so in his movie, he demeans scientists who don't share his views by referring to them as *"so-called"* skeptics.

Mind Control: Gore's introduction of Nazi Germany segues to the infamous propagandist, Joseph Goebbels (No, I'm not associating Gore with Goebbels, just that they both made a living from spewing propaganda). Goebbels was Hitler's Propaganda Minister: he developed *Principles of Propaganda* that included the following:

"Propaganda must label events and people with distinctive phrases or slogans; they (the phrases and slogans) must be capable of being easily learned, and they must be utilized again and again." Sound familiar?

Propagandists and politicians are familiar with the *illusory truth effect*, a cognitive bias whereby people equate repetition with truth. Goebbels' *"distinctive phrases or slogans"* are now called *talking points*.

And to ensure that the public doesn't inure to the message, AGW activists continually up the ante: From Gore's 2006 mild attack of calling climate change skeptics *"flat earthers"* who believe the Moon landings were staged (Note, Saul Alinsky's *Rules for Radicals*, item 5 says *"Ridicule is man's most potent weapon"*), to UN Secretary General, Ban Ki-Moon's statement to the BBC in 2007 that *"this year global warming poses as much a threat to the world as war;"* to U.S. Secretary of State John Kerry, speaking in Vienna, July 2016, at an international climate change conference, *"climate change is as dangerous as, if not more than, the threats posed by the Islamic State and other extremist groups."* Or how about David Roberts in *Grist Magazine* 2006, *"It's about the climate-change "denial industry" ...we should have war crimes trials for these bastards – some sort of climate Nuremberg."* Again, the thematic association with Nazi Germany but this time with the egregious implication that those "found guilty" of skepticism should be executed. From Nazi Germany to ISIS, climate change skeptics are associated with evil.

More examples of the venom: British journalist, George Monbiot,

"... *every time someone dies as a result of floods in Bangladesh, an airline executive should be dragged out of his office and drowned.*" Really, advocating murder? This is a modern version of "*burn the witches.*"

And Rajendra Pachauri, chairman of the IPCC, denigrating those who would stray from climate change orthodoxy, saying,

"*They are people who deny the link between smoking and cancer; they are people who say that asbestos is as good as talcum powder – I hope that they apply it to their faces every day.*" (*May 3, 2010, C2C Journal*). Quoted in an article by Professor Patrick Keeney, in which Dr. Keeney was calling for toning down the rhetoric. Unbelievable! Notice the straw man he introduces to associate skeptics with cancer.

The laypersons' understanding of climate change/global warming is to a large degree based on Al Gore's *An Inconvenient Truth (AIT)*, which misrepresented many *facts*. In October 2006, Michael Burton, a UK High Court Judge ruling on the admissibility of the film to be included in school curriculums (which in itself is a disgraceful implementation of brainwashing children, given the uncertainty of the CO_2 hypothesis), concluded that "*An Inconvenient Truth* **contains nine key scientific errors.**" Other sources have identified 35 scientific errors, noting that all the errors fall in the direction of supporting *the cause*, the statistically likelihood of which is less than 1 in 34 billion! (newsbusters.org/blogs/nb/noel-sheppard/2007/10/21/35-errors-discovered-al-gores-film).

Judge Burton ruled that errors had arisen "*in the context of alarmism and exaggeration in order to support Mr. Gore's thesis on global warming.*" It appears that AIT was intended as a political manifesto appealing to the fears of the average person and, most perniciously, that it was designed to create a *significant emotional event*.

A significant emotional event (SEE) can change perspectives, outlooks, actions and thoughts, and strong, vivid SEEs, as presented in AIT, can cause a degree of trauma, the resolution of which is provided by following the path prescribed by the guru, in this case, Al Gore. This is a form of the *Hegelian Dialectic*: create fear, the more the better, and then provide a solution, the outcome of which is usually loss of liberty and control of one's destiny to an elitist higher power. In the case of climate change, the higher power is the United Nations and its IPCC proxy.

In the most reductive sense, the fundamental basis of Hegelianism is that the human mind finds it difficult to grasp anything unless it can be split into binary opposites, i.e. good or evil; right or wrong, and so on. *Good* people support the IPCC agenda; those who don't support it are evil, they are killing the planet. And those good people are right (after all they are with 97% of scientists); skeptics are wrong. Once people are convinced they are good and right, they are ripe subjects for the propaganda techniques discussed in this chapter; and it works!

Politicians have used this idea for decades. In Oct 2005, Niki Raapana and Nordica Friedrich (crossroad.to/articles2/05/dialectic.htm) wrote,

*"**The Hegelian dialectic is the framework for guiding our thoughts and actions into conflicts that lead us to a <u>predetermined solution.</u>** Hegel's dialectic is the tool which manipulates us into a frenzied circular pattern of thought and action.... Hegelian conflicts steer every political arena on the planet, from the United Nations to the major American political parties, all the way down to local school boards and community councils. **Dialogues and consensus-building are primary tools of the dialectic,** and **terror and intimidation are also acceptable formats for obtaining the goal.**"* The *"pre-*

determined solution" is that human-caused CO_2 is responsible for climate change.

Terror is caused in large part by an intense fear; fear that we or our children's survival is at risk could certainly induce a sense of terror. Survival is the basic motivation in Maslow's hierarchy of needs; we must satisfy our survival needs before we concern ourselves with anything else; i.e., we must stop global warming!

Fear: The most powerful emotion is *fear*. Fear is a primal instinct that served us as cave dwellers and does so today; it often keeps us alive. But it is also exploited to convince people to behave in certain ways. According to Steven Hassan (a prolific author and leading authority on cult abduction and mind control), when asked how cult recruiters are able to convince highly intelligent, well-educated, free citizens to leave the comfort of their friends and family to work like slaves for a cult leader, responded,

"It is done through fear. They intentionally create phobias in the minds of their targets. Having done so, they can make them believe anything."

Marketers use fear to sell products or services. They present a scenario they hope will invoke a sense of fear. Then they offer a solution that entails using their product or service. It's a form of the Hegelian Dialectic. Consider the TV ads for buying gold or silver; they present a message that the financial world will collapse and our money will be worthless – that's scary! But, if we buy their gold or silver our savings and our families will survive. Or, *your child will get cervical cancer if she doesn't get the HPV vaccine,* and so on.

The British philosopher Bertrand Russell, once wrote *"neither a man nor a crowd nor a nation can be trusted... to think sanely under the influence of a great fear."* Machiavelli notoriously argued that a good leader should induce fear in the populace to control them (the rabble). Thomas Hobbes in *The Leviathan*, argued that fear effectively motivates the creation of a social contract in which citizens cede their freedoms to the sovereign. Strangely, Hobbes thought that was ok.

Fear is used to sell virtually everything, products, services, excessive government control, etc., and it's being used to sell global warming alarmism. In this endeavor, politicians and governments are abetted by biased media.

Media and Sensationalism: In 2002, J. Crib, an experienced editor, wrote in the Australasian Science Journal,

*"**The publication of "bad news" is** not a journalistic vice; it's clear instruction from the market; it's **what consumers**, on average, **demand**.... As a newspaper editor I knew, as most editors know, that if you print a lot of good news, people stop buying your newspaper. Conversely, **if you publish the correct mix of gloom, doom, and disaster, your circulation swells**. I have done the experiment."* This should tell us all we need to know to be skeptical of media sponsored hysteria.

In 2006, Time magazine's cover was *"Be Worried, be Very Worried – A Special Report on Global Warming."* Great headlines, captivating scare tactics.

*"**Entire nations could be wiped off the face of the Earth** by rising sea levels **if global warming is not reversed by the year 2000**. Coastal flooding and crop failures would create an exodus of "eco-refugees," threatening political chaos.... **governments have a 10-year window** of opportunity **to solve the greenhouse effect**."* (Brown, Director of the New York office of the UN Environment Program; Associated Press, June 30, 1989). Note, there's never a follow up report in the media, and it's been almost 30 years.

"By 1995, the greenhouse effect would be desolating the heartlands of North America and Eurasia with horrific drought, causing crop failures and food riots....By 1996 the Platte River of Nebraska would be dry, *while a continent-wide black blizzard of prairie topsoil will* stop traffic on interstates, strip paint from houses *and shut down computers....The Mexican police will round up illegal American migrants surging into Mexico seeking work as field hands."* (1990, Michael Oppenheimer from *Dead Heat: The Race against the Greenhouse Effect*).

Oppenheimer is a climate scientist and Albert G. Milbank Professor of Geosciences and International Affairs at Princeton University, but not much of a forecaster.

Just for fun, I researched data from the U.S. Census Bureau and from 1990 through 2016: just counting the top 20 U.S. states, approximately 39 million immigrants have entered the U.S.A, of which about 45% are of Hispanic origin. So much for the southern exodus! And incidentally, if the effects of CO_2 are global, why would Mexico fare better than the U.S.? And I don't remember the *"food riots,"* or the *"black blizzard of prairie top-soils shutting down computers."* Such nonsense, but never retractions.

Stephen Schneider, a prominent and respected climatologist and leading proponent of the AGW hypothesis, in a 1988 interview with *Discover* magazine said,

*"And like most people we'd like to see the world a better place, which in this context translates into our working to reduce the risk of potentially disastrous climate change. To do that we need to get some broad based support, to capture the public's imagination. That, of course, means getting loads of media coverage. **So we have to offer up scary scenarios, make simplified, dramatic statements, and make little mention of any doubts we might have.**"*

Dr. Schneider's intentions may be noble, to see the world a better place, but his methods are not. For ardent believers, the end justifies the means.

And from Senator Daniel Moynihan (1969),

"It is now pretty clearly agreed upon that the CO_2 content (in the atmosphere) will rise 25 percent by 2000. This could increase the average temperature near the Earth's surface by 7 degrees Fahrenheit. This in turn could raise the level of the sea by 10 feet. Goodbye, New York, Goodbye Washington, for that matter."

Based on the Mauna Loa CO_2 profile shown in Chapter 3 (Figure 3-7), he was fairly accurate in his prediction of CO_2 levels increasing by about 25%. It is after all increasing linearly. His other prediction however, based on a causal and dominant relationship between CO_2 and temperature was way off the mark; rather than an increase of 7°F (3.9°C), temperatures increased by less than 2°F (1.1°C), and of course sea levels did not increase by 10 feet. As with most politicians, Moynihan carefully crafted his comments with plausible deniability by inserting the word *"could"* in his predictions. And again, no follow-up retraction.

Here's what I find startling about the media: according to a Center for Disease Control article, March 2017, *smoking* is the leading cause of preventable death. The article lists the following points:

Worldwide, tobacco use causes nearly 6 million deaths per year, and current trends show that tobacco use will cause more than 8 million deaths annually by 2030. Cigarette smoking is responsible for more than 480,000 deaths per year in the United States. **That is about 1,300 deaths every day in the USA alone, equivalent to four B-747 Jumbo jet crashes PER DAY with 100% fatality**.

Where is the media hysteria? And this isn't in dispute, these are facts. Not even the most ardent climate change advocates claim such carnage under their worst-case scenario. Where is the Gore movie showing people suffering from lung disease? In his AIT movie, he does relate those who denied a connection between smoking and cancer with climate change skeptics – too great an opportunity to miss.

And, what is the government response? Make the tobacco companies put a warning on cigarette packages, add some tax, and ban advertising. This is an Inconvenient Truth!

Global warming and concern for the environment however, are mere proxies for ideologues to achieve their long-term objectives of a new world order one that at its core is a centralized socialist government under the auspices of the United Nations. They merely adopted *environmentalism* as the banner under which they could unite the masses. Is this hyperbole on my part? Well, consider the following:

"The objective, clearly enunciated by the leaders of UNCED (United Nations Conference on Environment and Development), is to bring about a change in the present system of independent nations. The future is to be World Government with central planning by the United Nations. Fear of environmental crises - whether real or not - is expected to lead to – compliance" **Dixy Lee Ray, former liberal Democrat governor of State of Washington, U.S.** (2015, *C3 Headlines*).

Remember the discussion on "fear?" And in true Hegelian style, truth doesn't matter, *"whether real or not"* - the end justifies the means.

"No matter if the science is all phony, there are collateral environmental benefits.... Climate change provides the greatest chance to bring about justice and equality in the world." (1988, Christine Stewart: Canadian Minister of the Environment from

1977-99). Again, truth doesn't matter, *"No matter if the science is all phony"* - the end justifies the means.

And while politicians and media continue to beat the environment drum for *saving the planet*, the ultimate objective, a world socialist government, was enunciated most clearly by Ottmar Edenhofer, IPCC chair from 2008 to 2015 who, in a 2010 interview jettisoned the environment mask, saying,

"One has to free oneself from the illusion that international climate policy is environmental policy. This has almost nothing to do with the environmental policy anymore... We redistribute de facto the world's wealth by climate policy."

Of course, it was never about the environment. How many AGW supporters would remain so ardent in their belief if they were aware of the ideologues long-term vision? In the words of H.L. Mencken,

"The urge to save humanity is almost always only a false face for the urge to rule it."

Science skeptics have always existed; they have been a healthy and essential feature of scientific research that has enhanced our knowledge of complicated subjects where we no longer blame thunderstorms on Thor, or bad harvests on unfortunate witches. But since climate change became a pseudo-secular religion it treats skeptics as heretics, with an intolerance not seen since the pre-enlightenment era.

The solar system is not geocentric, the Earth's rotation around the Sun isn't spherical, gravity isn't transmitted instantaneously across the cosmos, and General Relativity didn't need a cosmological constant to *prevent* the Universe from expanding. Aristotle and Ptolemy were wrong in their **geocentric view of the**

solar system that was supported by consensus for nearly 2000 years, and scientific errors were made by Copernicus, Newton, and Einstein. And, in the cases of the latter three scientists, who were skeptics in their day, their scientific theories were based on repeated verification of results and were accepted by the scientific community from decades to hundreds of years; but only until they weren't. So, Newton and Einstein can be wrong, but Gore, Kerry, et al, can't be?

In accordance with the scientific method, their theories were repeatable, testable, and falsifiable. When observations were inconsistent with the prevailing theory, scientists became skeptical of the theory and developed new or modified theories. Always in modern science, when a theory fails an experimental test it is abandoned. Albert Einstein said, *"No amount of experimentation can ever prove me right; a single experiment can prove me wrong."* Never was the science settled for those icons or in any other legitimate field of science; but for AGW activists, the science *is* settled and those who don't acquiesce to this dogma are *deniers;* they are *heretics;* they are evil, and they must be silenced.

Skepticism is the cornerstone of scientific inquiry except when dealing with climate change, when it is maligned. According to Professor Ian Plimer (2009, *Heaven and Earth: global warming the missing science*), *"unless dogma and orthodoxy are challenged, we retreat into a world of superstition and/or authoritarianism."* Furthermore, an important point missed in this *debate* is that **the onus of proof of a hypothesis rests on proposers of the hypothesis, not skeptics.**

How ridiculous to embark on major political policy changes and world treaties, spending hundreds of billions, potentially trillions of dollars (that we don't have) on battling a condition that, regardless of media bias to the contrary, is in dispute. It always comes down to *follow the money.* Hysteria is the mother's milk for politicians and the media – it secures government-provided mega funds without serious opposition, and it sells. Al

Gore, a major demagogue in promoting the global warming hysteria, is now worth more than $100 million and is a recipient of a Nobel Peace prize. His *An Inconvenient Truth* is a motion picture version of the *Book of Revelation* with its apocalyptic fear mongering. And there's nothing quite like an *end of the world* scare tactic to get the juices flowing. History is replete with examples of insane behavior when people are scared, from the slaughter of innocent Jews during the Middle Ages Plague (Black Death), to *witch trials.* Incidentally, according to University of Chicago Professor Emily Oster et al, (Winter 2004; *The Journal of Economic Perspectives*), the most active period of witchcraft trials in Europe coincided with the 400-year Little Ice Age when witches were blamed for poor harvests, etc.

97% Consensus: I'll begin by saying that of all the literature I have read on climate change, *more than* 97% of the authors agree that human-caused CO_2 has contributed to an increase in greenhouse gases and global warming; they just don't agree that it is the dominant force.

Publicists at the United Nations, politicians and their acolytes frequently claim that *only a few skeptics remain* – those who are still unconvinced about the existence of a catastrophic human-caused global warming emergency - and that there is a 97% consensus that human caused CO_2 is responsible for global warming and the impending catastrophe. These claims are not consistent with my research: following is an extract from the *Global Warming Petition Project*, published in 2008, signed by 31,487 American scientists, including 9,029 PhDs,

"There is no convincing scientific evidence that human release of carbon dioxide, methane, or other greenhouse gases is causing or will, in the foreseeable future, cause catastrophic heating of the Earth's atmosphere and disruption of the Earth's climate. Moreover, there is substantial scientific evidence that increases in atmospheric carbon dioxide produce

many beneficial effects upon the natural plant and animal environments of the Earth."

The purpose of the Petition Project was to demonstrate that the claim of *settled science* and an overwhelming *consensus* in favor of the hypothesis of human-caused global warming and consequent climate catastrophe is patently false. It is evident that 31,487 Americans with university degrees in science are not *a few*. If the 97% number was correct it would mean that more than *one million* scientists in America alone are in agreement with the IPCC charter which is, of course, absurd.

Dr. Roy Spencer, testifying at a Senate hearing in 2013, debunked the notion of the 97% consensus claims of the IPCC, politicians, media et al, saying that the number includes him and other colleagues because it includes people who think humans have *some influence* on the climate. It doesn't differentiate those, such as Dr. Spencer, who agree that humans have an effect on the climate but disagree that the effect is significant in comparison with natural causes. Dr. Spencer speaks as an expert on this subject; he is a climatologist, Principal Research Scientist at the University of Alabama, Huntsville, and was the U.S. Science Team leader from June 2002 to early October 2011 for the Advanced Microwave Scanning Radiometer for the Earth Observing System (AMSR-E) on the Aqua satellite which provided data about the Earth's water cycle among which was Arctic and Antarctic sea ice coverage and snow cover.

Legates et al (2014; *Climate Consensus and Misinformation*), demonstrated that only 0.5% of the abstracts of 11,944 scientific papers on climate-related topics published over 21 years from 1991-2011 had explicitly stated an opinion that supports the IPCC position.

Furthermore, according to the website notrickszone.com/skeptic-papers, in 2016, 500 peer-reviewed scientific papers published in scholarly journals strongly suggest that natural factors exert a significant or dominant influence on weather and climate, and that an anthropogenic CO_2 influence may be much more difficult to detect in the context of such large natural variability. The trend continued into 2017; in January alone, there were at least 17 papers published in scientific journals documenting that modern warming is not global, unprecedented, or remarkable. And, according to Kenneth Richard (2018, climatism.blog), in just the first 8 weeks of 2018, 97 scientific papers were published that cast doubt on the position that anthropogenic CO_2 emissions are the main cause of climate change.

And, the above refutation of the consensus claims notwithstanding, **consensus is not a scientific principle,** it is a political ploy (a primary tool of the Hegelian Dialectic) to end debate; once the consensus card is played, anything that questions the popular paradigm can be dismissed without reason, because *skeptics are deniers.* Dr. Michael Crichton said the following:

"There is no such thing as consensus science. If it's consensus, it isn't science. If it's science, it isn't consensus."

*"Historically, the claim of **consensus has been the first refuge of scoundrels**; it is a way to avoid debate by claiming that the matter is already settled."*

"I would remind you to notice where the claim of consensus is invoked. ***Consensus is invoked only in situations where the science is not solid enough.*** *Nobody says the consensus of scientists agrees that $E=mc^2$. Nobody says the consensus is that the Sun is 93 million miles away..."*

For more than 2,000 years, from Aristotle to Copernicus, there was probably 100% consensus that the Sun orbited the Earth. Copernicus and later Galileo were skeptics, as were Newton and Einstein.

Chapter 1 Summary: **Climate science has become an oxymoron;** as practiced by the IPCC and advocated by politicians and media acolytes it is a political platform supported by talking points rhetoric and reinforced by polemic condemnation of non-co-opted scientists. This chapter addressed political and media manipulation, specifically how they shape the minds of an uninformed public by employing propaganda techniques to first create fear and then offering a solution, their ideology, to assuage such emotions.

People respond to sensationalism, and politicians and media take full advantage of this with hyperbole and in many cases misleading statements made, they claim, for the greater good of society; i.e., the end justifies the means.

Many AGW activist leaders are on record stating that their ultimate goal is a global central government and the eradication of capitalism. A major strategy to achieve their agenda is to control the world's energy, but under the illusion that it's about environmentalism; i.e., saving the planet.

Targets of ideologues are skeptics; i.e., anyone, no matter their qualifications, who disagrees with their *cause célèbre* are vilified and associated with evil; they are *deniers*.

The 97% consensus nonsense is debunked for the falsity that it is. All the scientists whose work I reviewed agree that the Earth is warming, that CO_2 is increasing, and that humans are contributing to the CO_2 increase. There is not however, consensus, or anywhere near consensus, that anthropogenic CO_2 is the dominating force of the warming trend. Claims of consensus are simply a manipulative tactic that relieves politicians and other advocates of *the cause* of the need to validate claims or have knowledge of science; they simply defer to the consensus, even though it is fabricated. And of course, if the science is settled there would be no reason to play the *consensus card*, and there would be no reason to continue spending billions of dollars on science that is settled.

Chapter 2
Deceit, Lies, and Corruption

Another Jefferson quote: "*My reading of history convinces me that most bad government results from too much government.*" And that's precisely what's happened with the establishment of the United Nations (UN) Intergovernmental Panel for Climate Change (IPCC) whose messages to the public are controlled by multi-government representatives.

This chapter presents a disturbing picture of the IPCC's behavior, and that of those individuals and organizations, including the once-revered NASA, which are of similar mind in that *the end justifies the means* as it relates to climate change and the elimination of fossil fuel.

The IPCC, which controls the narrative for climate change activists, was established, not to assess the scientific basis associated with climate change but according to its *Principles Governing IPCC Work*, its role is to "**assess**...*the scientific basis of risk of* **human-induced climate change**...." That is, as stated in the Introduction, before it began to compile its reports the IPCC took "*human-induced climate change*" as an *a priori* conclusion when at best it should have been a hypothesis subject to scientific inquiry. Its charter does not even allow for research to determine *if* human-induced CO_2 is a significant contributor to climate change, and if so to what degree.

According to scientific principles, it is necessary to start with a hypothesis. In this case it could be "anthropogenic CO_2 is the cause of global warming," or something similar Then, the null hypothesis, "anthropogenic CO_2 is *not* the cause of global warming" is tested. Only by the rejection of this null hypothesis could it be established that there is or could be a human influence, and even then, it is at some predetermined *level of probability;* NOT an indisputable fact. The IPCC ignores the scientific method and

jumps to the conclusion, it's human-caused CO_2; that's that, it's settled!

To date, the IPCC has published five Assessment Reports (AR): 1990- First AR (FAR); 1995-Second AR (SAR); 2001 – Third AR (TAR); 2005 – Fourth AR (AR4), and; 2013 – fifth AR (AR5). The sixth AR is scheduled to be published in 2018. Assessment reports comprise research results, projections, and mitigation strategies from three working groups, WG1, WG2, and WG3. It is however, the Summary for Policymakers (SPM) that is generally used by politicians and advocates of *the cause* and, as demonstrated below, the SPM does not always reflect the scientific findings of the working groups.

The IPCC's published process for approving scientific articles has a penultimate review and approval by a panel of government representatives; no chance of political bias or influence on selected articles for final release there. From IPCC procedures, section 4,

"...*changes made* after acceptance by the working group (the scientists) *or the Panel* (politicians) **shall be those necessary to ensure consistency with the Summary for Policy Makers** or the Overview Chapter." This should be read very carefully for it is saying that scientific results are subject to government approval – Orwell's Ministry of Truth!

Since the SPM, the part of the report released to the public and upon which policy is based, is written by a political committee, and the IPCC charter is clear in its position that approved publications support its *a priori* conclusions, it often results in scientific studies being altered.

Scientific Working Group reports are often altered to be *"consistent with the Summary of Policy Makers"* which of course, makes no sense, except to ideologues. Surely, as a *summary* the SPM should be altered to be consistent with the scientific reports.

Numerous scientists have resigned from or simply refused to participate in IPCC working groups because of **the political process that constantly ignores input that refute its claims and alters scientific data** (after releasing the Summary Report) to support its agenda.

Dr. Chris Landsea, Hurricane Researcher at the National Oceanic and Atmospheric Administration's (NOAA) Atlantic Oceanographic and Metrological Laboratory, resigned from the IPCC when his contribution *"Little if any increase in hurricane strength in the next 80 years"* was replaced with different conclusions. In his resignation he wrote,

"I cannot in good faith continue to contribute to a process that I view as both being motivated by pre-conceived agendas and being scientifically unsound."

Dutch Professor Richard Tol resigned from the IPCC in 2014 because of the biased negative conclusions of the AR5 climate report. He said,

"The consequences of climate change are being systematically over-estimated.... The (IPCC) Panel is directed from within the environment lobby and not from within the science."

Dr. Richard Lindzen, Professor of Meteorology at MIT, was a lead author on Chapter 7 of the IPCC Third Assessment Report, published in 2001. In May of that year he was critical of the Summary for Policymakers, which he said misrepresents what scientists say. He also said that the IPCC encourages misuse of the Summary; that the Summary does not reflect the full document, and that the final version was modified from the draft in a way to exaggerate man-made warming. He did not participate in any later IPCC reports and has since been an outspoken critic of the IPCC and its process.

Examples of manipulative behavior are presented in Singer and Avery (2007, *Unstoppable Global Warming*), and Tim Ball's (2014, *The Deliberate Corruption of Climate Science*): they say that the IPCC's 1995 (Second Assessment Report) scientific reports and the Summary for Policy Makers were reviewed and approved by its scientists in December 1995. After the printed report appeared, six months later, in May 1996, the scientific reviewers discovered that major changes, deletions and additions, had been made, after they has signed off on the science chapters contents.

Deletions from the "approved" reports included the following:

"None of the studies cited above has shown clear evidence that we can attribute the observed climate changes to the specific cause of increases in greenhouse gases."

"… no study to date has positively attributed all or part of the climate change observed to man-made causes. Nor has any study quantified the magnitude of a greenhouse gas effect…in the observed data – an issue of primary relevance to policy makers."

"Any claims of positive detection and attribution of significant climate change are likely to remain controversial until uncertainties in the total natural variability of the climate system are reduced."

Contrast those statements with what was added to the "approved" reports:

"There is evidence of an emerging pattern of climate response to forcing by greenhouse gases and sulfate aerosols…. these results point toward a human influence on global climate."

"The body of statistical evidence in chapter 8 (the scientific reports), when examined in the context of our physical understanding of the

climate system, now points to a discernible human influence on the global climate."

In other words, after the SPM was released, the scientific research report, Chapter 8, was *corrected* to ensure consistency with the IPCC Charter. Dr. Frederick Saitz was an American physicist and a pioneer of solid state physics; he was president of Rockefeller University, and president of the United States National Academy of Sciences from 1962–1969, and was the recipient of the National Medal of Science, NASA's Distinguished Public Service Award, and other honors. He wrote the following admonition in a *Wall Street Journal op-ed, 12 June 1996,*

*"In my more than 60 years as a member of the American scientific community, including service as president of both the National Academy of Sciences and the American Physical Society, **I have never witnessed a more disturbing corruption of the peer-review process** than the events that led to this IPCC report."*

And then, according to Ball (2014) and the website *prisonplanet.com/exclusive-lead-author-admits-deleting-inconvenient-opinions-from-ipcc-report.html,* Ben Santer, a climate researcher and lead IPCC author of Chapter 8 of the 1995 IPCC Working Group 1 report, **admitted** in December 2009 **that he deleted sections of the IPCC chapter which stated that humans were not responsible for climate change**. Of course by then, 14 years later, it didn't make any difference, battle lines were drawn and trenches dug, and it certainly didn't get published by main stream media as a retraction.

The UK House of Lords, Select Committee on Economic Affairs; *The Economics of Climate Change; Volume I:* report published 6 July 2006, drew its conclusions,

"Overall, we are concerned that the IPCC process could be improved by rethinking the role that government-nominated representatives play in

*the procedures, and by ensuring that the appointment of authors is above reproach. If scientists are charged with writing the main chapters, it seems to us they must be trusted to write the summaries of their chapters without intervention from others. Similarly, scientists should be appointed because of their scientific credentials, and not because they take one or other view in the climate debate. ... At the moment, **it seems to us that the emissions scenarios are influenced by political considerations**....."*

Note, the UK is an ardent supporter of the IPCC agenda; it is not a skeptical body, so it is reasonable to assume that this was a carefully crafted statement designed to air its concerns but with minimal criticism. The first sentence that the IPCC could be *"improved"* by *"rethinking"* and *"ensuring the appointment of authors is beyond reproach"* is a vaguely masked condemnation. The second sentence doesn't criticize the process at all; *"it seems to us..."* dissembles the main point, that politicians are overriding the scientists. The third sentence is non-controversial, but it does imply that some scientists are appointed for other than their credentials. The last sentence is where the meat of the statement resides, but again, it caveats its condemnation by *"it seems to us"* but the message is clear: the IPCC SPM is influenced by politicians in one of the most significant areas of the debate – emission scenarios, which drive policies.

An article by P. Gosselin (notrickszone.com/2012/05/09) on 9 May 2012, provides the following comments by Eminent German physicist and meteorologist and former advocate of *the cause*, Klaus-Eckert Puls,

*"Ten years ago I simply parroted what the IPCC told us. One day I started checking the facts and data – first I started with a sense of doubt but then I became outraged when I discovered that **much of what the IPCC and the media were telling us** was sheer nonsense and **was not***

even supported by any scientific facts and measurements.... The entire CO_2 debate is nonsense."

If someone so knowledgeable in physics and meteorology was so easily deceived, it is not surprising that the general public is too.

Because its charter is to focus on human-caused CO_2 climate change the IPCC has no *raison d'être* if climate change can be shown to be mostly natural and that anthropogenic CO_2 is not the primary cause. For the IPCC and its adherents the problem HAS to be human-caused CO_2, and only human-caused CO_2. Billions of dollars, power, prestige, celebrity, etc are enough incentive for climate change activists to maintain an intractable stance against scientific inquiry that might jeopardize their status quo. To quote a line from Al Gore's movie, *"It is difficult for a man to understand something, if his salary depends on his not understanding it."* I agree.

My skepticism was reinforced by learning of other serious data manipulations as evidenced by the infamous "Hockey Stick" ruse, data tampering by NASA and individual supporters of *the cause*, and the scandalous behavior exposed by "Climategate" of the tactics employed by AGW advocates and lead IPCC authors, et al.

Hockey Stick scam: The IPCCs First Assessment Report (FAR), published in 1990, included a graph of the past 1,000 years of global climate history showing a Medieval Warm Period with warmer temperatures than today (obviously before industrial-age CO_2), and a Little Ice Age with temperatures lower than those recently experienced. The original version of the graph, shown as Figure 2-1, was produced by H.H. Lamb (1965, *The Early Medieval Warm Epoch and its Sequel*). Note, the vertical axis, Temperature Change, is a nominal representation with no temperature dimensions.

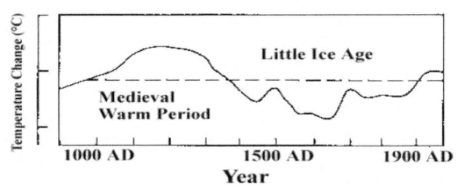

Figure 2-1 Medieval Warm Period and Little Ice Age versus Time

The FAR showed both the Medieval Warm Period and the Little Ice Age (Lamb data) with added dimensions and some additional "smoothing" as shown in Figure 2-2.

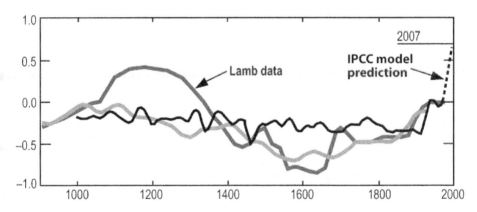

Figure 2-2 IPCC (First Assessment Report) – Temperatures over past Millennia

Since this was published in 1990, the curve from 1990 to 2007 must have been based on model predictions, and clearly not based on mathematical curve fitting that would not result in the explosive rise from the late 1990s to 2007. However, as disclosed by Climategate emails discussed later, both of these graphs were unacceptable to IPCC et al zealots.

In any case, the IPCC Third Assessment Report (TAR), published 11 years, later presented a different version; gone were the Medieval Warming and the Little Ice Age. This allowed the revised graph to depict steady temperatures for a thousand years followed by almost exponential warming over the past century. The revised (Hockey Stick) graph is shown as Figure 2-3; much more acceptable to *the cause*!

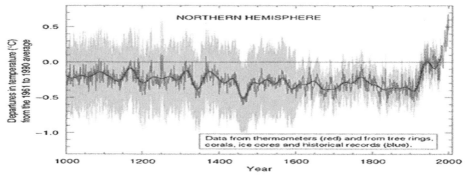

Figure 2-3 IPCC TAR Hockey Stick

The hockey stick profile is enthusiastically presented in most global-warming advocacy presentations. It appeared on the second page of the IPCC TAR Summary Report (used by government policy makers), intended to visually underscore their argument that human-caused CO_2 is the problem, and a serious problem.

Apart from the obvious deception (omission of the Medieval Warm Period), the hockey stick failed a fundamental scientific test, it wasn't reproducible; a critical requirement of the scientific method. Other scientists used the same data and procedures to try to reproduce the original findings. Steve McIntyre and Dr. Ross McKitrick attempted but failed to reproduce the findings and claimed a misuse of data and statistical techniques. In effect, they reported, the hockey stick was meaningless. This created a dispute with the hockey stick authors (Michael Mann, et al). The National Academy of Sciences appointed a committee, chaired by Professor Wegman PhD, a professor of statistics at George Washington University, to investigate and arbitrate. His committee found in favor of McIntyre and McKitrick (MM) as follows:

".... Because of lack of full documentation of their (Mann et al) *data and computer code,* **we have not been able to reproduce their research.** *We did, however, successfully recapture similar results to those of MM.* **This recreation supports the critique of the MBH98 methods.***"*

Note: MBH98 refers to the initial hockey stick reconstruction method used by Mann et al. In other words, **the hockey stick scenario is false.**

The deceptive ruse was exposed, but the hockey stick graph is still widely used by advocates. But then, according to Paul Watson, Canadian environmental activist and founder of the Sea Shepherd Society, and a board member of Greenpeace, *"It doesn't matter what is true; it only matters what people believe is true."*

More deceit,

"In 1995, I published a short paper in the academic journal "Science." In that study, I reviewed how borehole temperature data recorded a warming of about one degree Celsius in North America over the last 100-150 years. The week the article appeared, I was contacted by a reporter for National Public Radio. **He offered to interview me, but only if I would state that the warming was due to human activity.** *When I refused to do so, he hung up on me. I had another interesting experience around the time my paper in Science was published. I received an astonishing email from a major researcher in the area of climate change. He said,* **we have to get rid of the Medieval Warm period.***'"* (David Deming, University of Oklahoma, before the Senate Environment and Public Works Committee, December 6, 2006).

Articles of Deception: In a major paper published in the British science journal *Nature* (4 July, 1996, Vol.382, p.39-46) many of the top players in the greenhouse industry, including Benjamin Santer, lead author IPCC Second Assessment Report (SAR), Chapter 8 of Working Group 1 report; Tom Wigley of National Center for Atmospheric Research; Philip Jones of the Climate Research Unit (CRU); John Mitchell of the U.K. Hadley Centre; A. Oort and R. Stouffer of NOAA/Geological Fluid Dynamics Laboratory, et al, lent their names to a paper titled *A Search for Human Influences On The Thermal Structure Of The Atmosphere.* In their article, they presented the graph shown as Figure 2-4 indicating a significant increasing temperature trend from 1963 through 1987, a trend they asserted strongly correlated with IPCC models. They claimed they had validated the models, which by extension, meant climate model projections were reliable.

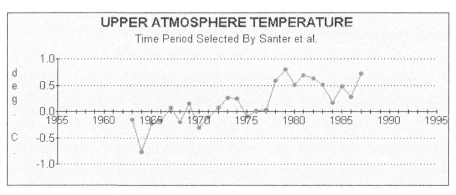

Figure 2-4 "Proof" of Human Influence on Upper Atmosphere Temperature

(Source: *John Daly blog*)

This paper was trumpeted by AGW advocates as proof of their cause and validation of the models by matching them with empirical, observed data. However, errors (of omission) were spotted quickly by John Daly (john_daly.com/sonde.htm). Recognizing that the Santer et al graph displayed only a partial representation of available data, created the graph shown as

Figure 2-5 that included the entire period of performance represented by the data. It is from the same source database from which Figure 2-4 was derived but reflects the entire data sample.

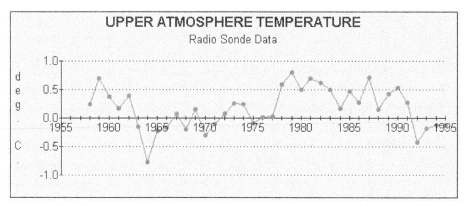

Figure 2-5 The Complete Dataset (Source: *John Daly blog*)

According to Daly, Santer et al chose the dates in Figure 2-4 as a basis on which to compare observed conditions against those that the models would predict. When the full dataset was shown (*Nature, vol. 384, 12 Dec 96*) five months after the Santer *eureka* moment, it was clear that Santer's version reflected only a subset of the available data. This in my view was tantamount to fraud. Remember the caution from the Preface regarding cherry-picking timeframes? If the decade 1979 to 1989 was chosen as the timeframe from the same dataset it would show a dramatic cooling phase. But that would also be wrong. What we need for a meaningful debate on climate change is scientific honesty and integrity. Unfortunately, Santer's depiction of the data, despite being shown to be manipulated by the omission of valid probative evidence, has become the mantra for AGW advocates and claims that the models represent reality.

Although, as stated earlier in this section, Santer later admitted that he deleted sections of the IPCC chapter which stated that humans were not responsible for climate change, by then the objective of getting the *"discernible human influence"* on the world stage had been achieved and etched on tablets. In any case, while Figure 2-4, the Santer graph, claimed (falsely) to prove the efficacy of the climate models it did nothing to support the argument that anthropogenic CO_2 is the driving force of global warming – it shows temperature oscillations ranging from approximately - 0.8°C to +0.8°C while CO_2 steadily increased over that period. As will be shown in the section on Climategate, this lack of correlation between CO_2 and temperature is the Achilles heel for AGW alarmists.

Data Manipulation: In 1981, Dr. James Hansen, considered by many to be the *godfather* of global warming advocacy, was the Director of the NASA Goddard Institute for Space Studies. He was also the lead author of a paper published in the prestigious journal *Science* entitled *"Climate Impact of Increasing Atmospheric Carbon Dioxide."*

In the paper, Hansen and his colleagues reported (and illustrated with multiple graphs) the widely accepted 100-year (1880-1980) record of hemispheric and global temperature changes. At the time, most climate scientists were reporting that Northern Hemisphere's (N-H) temperatures had undergone a rapid warming of between +0.8 and +1.0°C between the 1880s and 1940. Then, after 1940 and through 1970, N-H temperatures were reported to have dropped by about -0.5 to -0.6°C, a decades-long cooling trend which at the time fomented widespread debate about global *cooling* in the scientific community (see Chapter 3, opening section). Like their peers, and in accordance with the data, NASA's Hansen and his co-authors reflected these data, showing that the N-H had warmed by about 0.8°C between the 1880s and 1940, and then cooled by approximately 0.5°C between 1940 and 1970.

A graph of "Observed temperature" for the N-H was included in the 1981 paper to illustrate these climatic trends (Figure 2-6). The N-H varied from a low of -0.4°C in the 1880s to +0.4°C in 1940 (an increase of 0.8°C). In their paper, the authors showed a similar temperature pattern for the Southern Hemisphere (S-H), but a less severe variance (the N-H has far more land than the S-H which is mostly ocean and slow to change temperature).

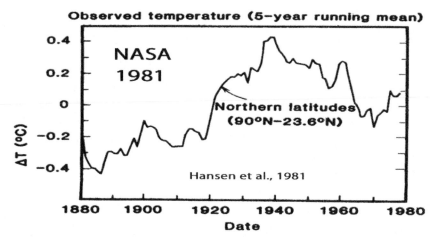

Figure 2-6 Observed Temperature from 1880 through 1980

Dr. Gavin Schmidt, Director of NASA Goddard Institute for Space Studies (GISS) after Hansen retired in 2013 presented a completely different version of the N-H temperature record for the hundred years 1880 to 1980. Instead of leaving the historically observed temperatures as they were recorded, NASA *invented* new ways to portray pre-1981 temperature history. Figure 2-7 shows Hansen's original curve (top dark line reflecting the historical data – NH Annual Mean 1981) overlaying NASA's revised curve (bottom – NH Annual Mean 2107). Remember, the IPCC was formed in 1988 – prior to then there were concerns, as expressed in Hansen et al article, but there was not yet a *cause*. Now there is!

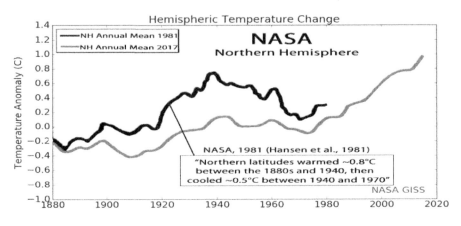

Figure 2-7 NASA's New Temperature Trend (2017)
versus Original Observations (1981)

NASA reduced the 1880s to 1940s warming from 0.8°C to about 0.4°C, and warmed up the three decades of NH cooling (1940 - 1970) by about +0.3°C (from -0.5°C to -0.2°C). In this way, the overall temperature from the 1880s to 2017 appears more linear and more in line with the trend in atmospheric CO_2 (linear continuous upward trend). **When the facts didn't fit the models, NASA changed the facts.**

If the originally recorded observations for the N-H had *not* been altered from the temperature record, the pre-1981 trend would look like it does in Figures 2-6 or the top, thick black line of Figure 2-7; i.e., if the data had not been tampered with, it would be clear the N-H surface temperatures have undergone oscillations, or warming-cooling-warming cycles, with a rate of warming from about 1920 to 1934 similar to that from 1980 to present.

Why did NASA eliminate the early 20th century warming and mid-20th century cooling? According to Daly, the fundamental reason why NASA manipulated past temperature data was,

"So that the historical climate record may conform to the IPCC models that presume variations in surface temperatures are predominantly determined by anthropogenic CO_2 emissions."

NASA apparently was caught in a state of *cognitive dissonance* with the unacceptable recognition that (a) anthropogenic CO_2 emissions were rising slowly while surface temperatures were rising dramatically (1880s-1940s), and (b) surface temperatures were cooling significantly (1940s to 1970s) while anthropogenic CO_2 emissions continued its upward trend. These observations undermined the models and more importantly, *the cause.* NASA had uncovered an inconvenient truth that didn't support its agenda. So, to counteract this, it engaged in a decades-long effort to change historical temperature data; *the end justifies the means.*

Incidentally, an obvious anomaly that was not discussed in the Daly paper is that the NASA curve shown in Figure 2-7 indicates a steep continuous temperature increase from 1980 to present when it is known, even acknowledged by the IPCC as a *pause,* that there was a levelling of temperature from late 1990s through about 2014, after which warming continued. But those data also wouldn't support *the cause.*

NASA also engaged in another, what I consider to be a scandalous activity in support of *the cause*; this time, one of concealment. In 2005, NASA posted on the web a reasoned and thoughtful explanation of climate forcing addressing both natural causes and human CO_2. Its title was: *"What Are The Primary Forcings of The Earth System?"* And its explanation is as follows:

"The Sun is the primary forcing of Earth's climate system. Sunlight warms our world. Sunlight drives atmospheric and oceanic circulation patterns. Sunlight powers the process of photosynthesis that plants need to grow. Sunlight causes convection which carries warmth and water vapor up into the sky where clouds form and bring rain. In short, the Sun drives almost every aspect of our world's climate system and makes possible life as we know it.

Earth's orbit around, and orientation toward, the Sun changes over spans of many thousands of years. In turn, these changing "orbital mechanics" force climate to change because they change where and how much sunlight reaches Earth. Thus, changing Earth's exposure to sunlight, forces climate to change. According to scientists' models of Earth's orbit and orientation toward the Sun indicate that our world should be just beginning to enter a new period of cooling – perhaps the next ice age.

However, a new force for change has arisen: humans. After the industrial revolution, humans introduced increasing amounts of greenhouse gases into the atmosphere, and changed the surface of the landscape to an extent great enough to influence climate on local and global scales. By driving up carbon dioxide levels in the atmosphere (by about 30 percent), humans have increased its capacity to trap warmth near the surface.

Other important forcings of Earth's climate system include such "variables" as clouds, airborne particulate matter, and surface brightness. Each of these varying features of Earth's environment has the capacity to exceed the warming influence of greenhouse gases and cause our world to cool. For example, increased cloudiness would give more shade to the surface while reflecting more sunlight back to space. Increased airborne particles (or "aerosols") would scatter and reflect more sunlight back to space, thereby cooling the surface. Major volcanic eruptions (such as that of Mt. Pinatubo in 1992) can inject so much aerosol into the atmosphere that, as it spreads around the globe, it reduces sunlight and cause Earth to cool. Likewise, increasing the surface area of highly reflective surface types, such as ice sheets, reflects greater amounts of sunlight back to space and causes Earth to cool.

Scientists are using NASA satellites to monitor all of the aforementioned forcings of Earth's climate system to better understand how they are changing over time, and how any changes in them affect climate."

These excerpts, which capture what I have learned to be the essence of climate forcing, were provided by Jamie Spry in a June

2018 blog, (principia-scientific.org/the-orwellian-era-of-nasa-climate-pseudoscience). They are from a NASA webpage produced in 2005, (science.nasa.gov/.../what-are-the-primary-causes-of-the-Earth-system-variability). This was, however, withdrawn in 2010, presumably because it didn't support the fixation on CO_2 as the sole cause of climate change. I found that an attempt to view this page will receive the following notification, *"Access Denied – You are not authorized to view this page."* So how do we know it existed? Well, Spry provided the following screenshot:

Word for word from the NASA Science *Earth* article. What happened? Most likely politics; this was removed during President Obama's first term and, based on their various speeches and representations the President, his Secretary of State, John Kerry, et al, had consumed the "Kool-Aid." But to refuse access of a tax-funded article to the public, when there are clearly no national security implications – after all it was available for 5 years - is outrageous and demonstrates extreme government bias in controlling information.

Another example of data manipulation was exposed by the Science and Public Policy Institute (SPPI) in 2007. In his movie, *An Inconvenient Truth*, Al Gore posed in front of a giant representation of ice core *evidence*, shown as two graphs, one of temperature levels and the other of CO_2 levels, over a period of hundreds of thousands of years.

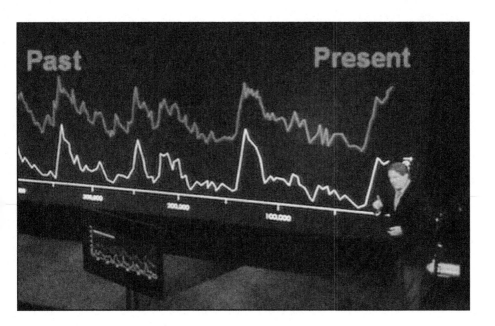

The relationship between CO_2 and temperature over the period is known by the scientific community to be that **CO_2 LAGS temperature**. Mr. Gore made an Oscar-winning performance deprecating those that can't see that *"when CO_2 increases, temperature increases,"* the implication being that CO_2 drives temperature, yet, in a 2007 (YouTube) appearance before Congress, in response to a statement by Joe Barton (Texas representative) he, **Gore, admitted that in the past *"CO_2 has LAGGED temperature."*** But that was before the industrial age following which, apparently, the laws of nature ceased to exist!

But more egregious than the Gore dissemblance was in a book by Laurie David and Cambria Gordon (2007, *The Down-To-Earth Guide to Global Warming*); it targeted school children and appears to be a blatant attempt to misinform (i.e., brainwash, re-educate) them. A 2007 article in the Business Wire by SPPI titled, *"SPPI Exposes Fundamental Scientific Error in Laurie David's 'Global Warming' Book for Children"* said the following:

"A fundamental scientific error lurks in a book calculated to terrify schoolchildren about 'global warming,' Robert Ferguson, SPPI president, announced today: 'The Down-To-Earth Guide to Global Warming,' by Laurie David and Cambria Gordon, **is intentionally designed to propagandize unsuspecting school children** *who do not have enough knowledge to know what is being done to them.*

They report that the book makes the claim,

"The more the carbon dioxide in the atmosphere, the higher the temperature climbed. The less carbon dioxide, the more the temperature fell. You can see this relationship for yourself by looking at the graph. What makes this graph so amazing is that by connecting rising CO_2 to rising temperature scientists have discovered the link between greenhouse-gas pollution and global warming."

This, which essentially parroted Gore's misrepresentation, was after all, the purpose of the piece; to convince children that humans were causing global warming.

The SPPI paper states,

"... in order to contrive a visual representation for their claim that CO_2 controls temperature change, the authors present unsuspecting children with an altered temperature and CO_2 graph that reverses the relationship found in the scientific literature."

The SPPI article shows the following:

A Fundamental Scientific Error in "global warming" Book for Children

On page 18 of Laurie David's new children's global warming book, there is a glaring scientific error.

David tells children:

Deep down in the Antarctic ice are atmosphere samples from the past, trapped in tiny air bubbles. These bubbles, formed when snowflakes fell on the ice, are the key to figuring out two things about climate history: what temperatures were in the past and which greenhouse gases were present in the atmosphere at that time.

The more the carbon dioxide in the atmosphere, the higher the temperature climbed. The less carbon dioxide, the more the temperature fell. You can see this relationship for yourself by looking at the graph:

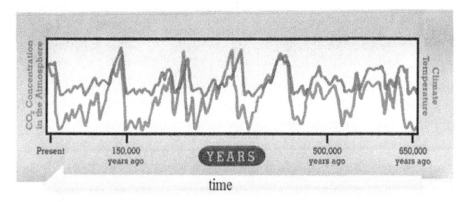

What makes this graph so amazing is that by connecting rising CO₂ to rising temperature scientists have discovered the link between greenhouse-gas pollution and global warming."

What **really** makes their graph "amazing" is that it's dead wrong. In order to contrive a visual representation for their false central claim that CO₂ controls temperature change, David and co-author Cambria Gordon present unsuspecting children with an altered temperature and CO₂ graph that falsely **reverses** the relationship found in the scientific literature.[1]

The last paragraph by SPPI exposes the lie. As stated in the caption, **the authors reversed the relationship from that recorded in the scientific literature; they mislabeled the curves to create the false impression that CO₂ leads temperature.** SPPI produced a corrected graph, in accordance with the scientific literature, shown below,

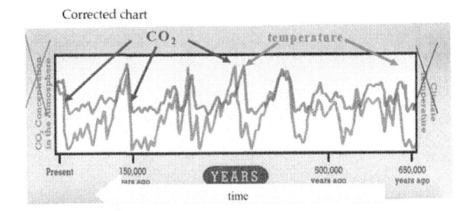

Corrected chart

For the full SPPI report, visit:
http://scienceandpublicpolicy.org/images/stories/papers/other/david_book.pdf

Once the curves are corrected (and read from right to left), it is obvious that over *geologic time* temperature rises before CO_2.

Climategate: On 19 November 2009, a story erupted on the web that seriously undermined the credibility of the AGW hypothesis and its advocates. Another batch was exposed two years later on 22 November 2011. Thousands of e-mails and other documents were disseminated via the Internet through the hacking of a server used by the Climatic Research Unit (CRU) of the University of East Anglia in Norwich, England. These e-mails have been interpreted by some as suggesting grave scientific misconduct and even conspiracy by leading climate scientists and IPCC authors and, as disclosed below, examination of revealed emails support these interpretations. At least three themes emerged from the released emails: (1) **prominent scientists** central to the global warming debate **concealed data** that did not advance their cause; (2) **these scientists** appear to **view global warming as a** *cause* rather than a balanced scientific inquiry, and; (3) **many of these scientists admit** to each other **that much of the science is weak and dependent on deliberate manipulation of facts and data.**

Starting with a copy of an email involving three of the authors associated with the *Nature* 1996 paper *"A Search for Human Influences on the Thermal Structure of the Atmosphere"* discussed earlier.

From: Tom Wigley <wigley@ucar.edu>
To: Phil Jones <p.jones@uea.ac.uk>
Subject: 1940s
Date: Sun, 27 Sep 2009 23:25:38 -0600
Cc: Ben Santer <santer1@llnl.gov>

<x-flowed>
Phil,

Here are some speculations on correcting SSTs to partly explain the 1940s warming blip.

If you look at the attached plot you will see that the land also shows the 1940s blip (as I'm sure you know).

So, if we could reduce the ocean blip by, say, 0.15 degC, then this would be significant for the global mean -- but we'd still have to explain the land blip.

I've chosen 0.15 here deliberately. This still leaves an ocean blip, and i think one needs to have some form of ocean blip to explain the land blip (via either some common forcing, or ocean forcing land, or vice versa, or all of these). When you look at other blips, the land blips are 1.5 to 2 times (roughly) the ocean blips -- higher sensitivity plus thermal inertia effects. My 0.15 adjustment leaves things consistent with this, so you can see where I am coming from.

Removing ENSO does not affect this.

It would be good to remove at least part of the 1940s blip, but we are still left with "why the blip".

Wigley and Jones were co-authors of the misleading paper *A Search for Human Influence on the Thermal Structure of the Atmosphere* discussed earlier in the section labeled Articles of Deception.

Dr. Tom Wigley served as Director of the Climatic Research Unit at the University of East Anglia from 1979 to 1993, and was a Senior Scientist at the National Center for Atmospheric Research from 1993 to 2006. He contributed to many IPCC reports and was a joint recipient of the 2007 Nobel Peace Prize that the IPCC shared with Al Gore.

Dr. Phil Jones, former Director of the CRU and Professor in the School of Environmental Sciences at the University of East Anglia, was a contributing author to Chapter 12, *Detection of Climate Change and Attribution of Causes*, of the IPCC Third Assessment Report (TAR) and a Coordinating Lead Author of Chapter 3, *Observations: Surface and Atmospheric Climate Change*, of the IPCC Fourth Assessment Report (AR4), and a central figure in the scandal wrote,

*"**One way to cover yourself** and all those working in AR5 **would be to delete all emails** at the end of the process."*

And, in another email, Jones wrote,

*"**Any work we have done in the past** is done on the back of the research grants we get – and **has to be well hidden**. I've discussed this with the main funder (U.S. Dept of Energy) in the past and **they are happy about not releasing the original station data**."*

The U.S. DOE a co-conspirator? President Eisenhower was extremely prescient in his farewell address. Jones wrote to a colleague, Penn State University scientist Michael E. Mann, (of Hockey Stick fame),

*"Mike, **can you delete any emails** you may have had with Keith [Briffa] re AR4 [UN Intergovernmental Panel on Climate Change 4th Assessment]? **Keith will do likewise**. ... **We will be getting Caspar [Ammann] to do likewise**. I see that CA [the Climate Audit Web site] claims they discovered the 1945 problem in the Nature paper!!"*

Deleting emails is the modern equivalent of burning books, a strategy invoked by dictators and tyrants throughout history.

Explanation of the comment referencing the 1945 problem: On the 30 May 2008, *Nature* reported the following:

"The time series of land and ocean temperature measurements, begun in 1860, shows a strange cooling of about 0.3 °C in the global mean temperature in 1945, relative to the 1961–90 average. The sharpness of the drop stands out even more if the signatures of internal climate variability, such as those associated with El Niño events, are filtered from the record."

It continues, *"This cooling at the end of the Second World War is one of several temperature drops in the record. But unlike others, such as the 1991 cooling caused by the eruption of Mount Pinatubo in the Philippines, it... is not associated with any known climatic or geological phenomenon.... "*

This is significant because once again it showed that they were aware of the Achilles heel, that there is no controlling relationship between CO_2 (which followed a steady increase over the entire 20th century) and temperature (which declined over this particular period).

Dr. Peter Thorne, Professor in Physical Geography (Climate Change) at Maynooth University in Ireland and Director of the Irish Climate Analysis and Research Unit S group (ICARUS), and a Lead Author on the IPCC Fifth Assessment Report (AR5), in one of the most damning of the emails wrote,

"Observations do not show rising temperatures throughout the tropical troposphere unless you accept one single study and discount a wealth of others. This is just downright dangerous. We need to communicate the uncertainty and be honest. Phil, hopefully we can find time to discuss these further if necessary. I also think the

science is being manipulated to put a political spin on it which for all our sakes might not be too clever in the long run."

This shows that some of the climate scientists were struggling with what they knew to be a conflict of scientific and personal integrity.

In an email to Michael Mann, Tom Wigley acknowledges the lies,

"Mike, the Figure you sent is very deceptive ... there have been a number of dishonest presentations of model results by individual authors and by IPCC."

In an attempt to intimidate a skeptic, Mann writes: *"I have been talking w/folks in the states about finding an investigative journalist* **to investigate and expose**" skeptic Steve McIntyre who, together with Dr. McKitrick successfully challenged and discredited Mann's hockey stick. But the intent of the language is appalling *"to investigate and expose."* Several hundred years earlier he could have hired Matthew Hopkins, the Witchfinder General!
The emails also reveal attempts to politicize the debate and advance predetermined outcomes.

"The trick may be to decide on the main message and use that to guide what's included and what is left out" (of IPCC reports), writes Jonathan Overpeck, a coordinating lead author for IPCC's climate assessments, AR4 and AR5.

Scientists advocating employing *tricks*! Why would they need to if their case was supportable?
I believe the point has been firmly established – Climategate showed a total disregard for adhering to scientific principles (as well as moral principles), it showed that those who are skeptical

should be "*exposed*," and it showed that where data didn't support its agenda IPCC scientists were prepared to conceal, alter, lie, and/or hide real data. In another email, Mann writes,

"*I gave up on [Georgia Institute of Technology climate professor] Judith Curry a while ago. I don't know what she thinks she's doing, but **it's not helping the cause**.*" *The cause?* Scientific inquiry doesn't have a cause.

Dr. Judith Curry was a supporter of the IPCC consensus on climate change until Climategate. After her denouncement of the scandalous behavior revealed by the leaked emails, she became a target of the *cabal*. It did not intimidate her into silence. Dr. Curry was Professor and former Chair of the School of Earth and Atmospheric Sciences at the Georgia Institute of Technology.

Dr. Curry (May 7, 2012, *The Legacy of Climategate*), discusses two articles written by Reiner Grundmann, (1) *Climategate and the Scientific Ethos*, and (2) *The Legacy of Climategate: Revitalizing or Undermining Climate Science and Policy?* Dr. Grundmann is Professor of Science and Technology Studies (STS) at the University of Nottingham (UK) and Director of its interdisciplinary STS Research Priority Group. Excerpts from (1) include the statement,

"*The exposed **climate scientists … gave preferential treatment to close allies**. They did not share their data …. They did not act in a disinterested way as the whole e-mail communication reveals. On the contrary, they acted strategically, showing self-interest and zeal. **Above all, they wanted to communicate the political message of their research** (that the Northern Hemisphere has never been as warm in the past millennium as it is at present) **and boost their own careers**. Finally, **they** did not foster organized skepticism but **tried to stifle skeptical voices**.*"

And from (2),

*"...what the emails reveal are problematic practices of **leading climate researchers acting as zealous gatekeepers** in a scientific and political project."*

On 13 Dec 2015, *The Corbett Report* published the following transcript of Dr. Judith Curry's testimony at a U.S. Senate Commerce Committee Hearing on *"Data or Dogma? Promoting Open Inquiry in the Debate over the Magnitude of the Human Impact on Earth's Climate."* Her presentation is compelling and should at the least inspire a little introspection by supporters of *the cause*. My comments, following some of the paragraphs, are not in italics.

*"Prior to 2009, I felt that supporting the IPCC consensus on climate change was the responsible thing to do. I bought into the argument: "Don't trust what one scientist says, trust what an international team of a thousand scientists has said, after years of careful deliberation." That all changed for me in November 2009, following the leaked Climategate emails, that illustrated the sausage making and even **bullying that went into building the consensus.***

Remember from Chapter 1, consensus building is a primary manipulation tool of the Hegelian Dialectic.

*I starting speaking out, saying that scientists needed to do better at making the data and supporting information publicly available, being more transparent about how they reached conclusions, doing a better job of assessing uncertainties, and actively engaging with scientists having minority perspectives. The response of my colleagues to this is summed up by the title of a 2010 article in the Scientific American: **Climate Heretic** Judith Curry Turns on Her Colleagues.*

This charge gets to the root of the problem we are dealing with in the climate debate: *heretic* is used almost exclusively in a religious context, someone who maintains views contrary to the established faith or belief. We don't hear this term used in any other field of science, because true science is not based on faith.

*I came to the growing realization that **I had fallen into the trap of groupthink. I had accepted the consensus based on 2nd order evidence:** the assertion that a consensus existed. I began making an independent assessment of topics in climate science that had the most relevance to policy.*

It's easy to see how people get caught up in the consensus nonsense when someone with Dr. Curry's credentials fell into the groupthink trap. Recall a similar reaction by Dr. Klaus-Eckert Puls, an eminent German physicist and meteorologist, who also *"parroted what the IPCC told us"* until he checked the facts and concluded that *"the entire CO_2 debate is nonsense."*

*What have I concluded from this assessment? Human caused climate change is a theory in which the basic mechanism is well understood, but whose magnitude is highly uncertain. No one questions that surface temperatures have increased overall since 1880, or that humans are adding carbon dioxide to the atmosphere, or that carbon dioxide and other greenhouse gases have a warming effect on the planet. **However there is considerable uncertainty and disagreement about the most consequential issues:** whether the warming has been dominated by human causes versus natural variability, how much the planet will warm in the 21st century, and whether warming is 'dangerous'.*

The central issue in the scientific debate on climate change is the extent to which the recent (and future) warming is caused by humans versus natural climate variability. *Research effort and funding has focused on understanding human causes of climate change.*

We have however, been misled in our quest to understand climate change, by not paying sufficient attention to natural causes of climate change, in particular from the Sun and from the long-term oscillations in ocean circulations.

The IPCC Charter focuses only on human-caused CO_2, yet recall from earlier in this chapter the NASA summary of *The Primary Forcings of the Earth System*, that began with the statement **"The Sun is the primary forcing of Earth's climate system."**

Why do scientists disagree about climate change? The historical data is sparse and inadequate. There is disagreement about the value of different classes of evidence, notably the value of global climate models. There is disagreement about the appropriate logical framework for linking and assessing the evidence. And scientists disagree over assessments of areas of ambiguity and ignorance.

How then, and why, have climate scientists come to a consensus about a very complex scientific problem that the scientists themselves acknowledge has substantial and fundamental uncertainties?

Climate scientists have become entangled in an acrimonious political debate that has polarized the scientific community. *As a result of my analyses that challenge IPCC conclusions, I have been called a denier by other climate scientists, and most recently by Senator Sheldon Whitehouse. My motives have been questioned by Representative Grijalva, in a recent letter sent to the President of Georgia Tech.*

It is outrageous that Dr. Curry's motives are maligned by politicians who have no knowledge of the science. Her views just don't support the cause – she is a *heretic*!

There is enormous pressure for climate scientists to conform to

the so-called consensus. This pressure comes not only from politicians, but from federal funding agencies, universities and professional societies, and scientists themselves who are green activists. **Reinforcing this consensus are strong monetary, reputational, and authority interests.**

Recall from the Introduction Chapter and Skinner Operant Conditioning, whereby behavior is modified by rewards and punishments.

In this politicized environment, advocating for CO_2 emissions reductions is becoming the default, expected position for climate scientists. This advocacy extends to the professional societies that publish journals and organize conferences. Policy advocacy, combined with understating the uncertainties, risks destroying science's reputation for honesty and objectivity – without which scientists become regarded as merely another lobbyist group."

One of Climategate's central figures, Phil Jones, appears to have had a conscience attack following the exposure of his and others actions revealed by the emails; according to a report in the Daily Telegraph, December 2017, *"Professor Jones, the scientist at the centre of the "Climategate" leaked email scandal, told how he considered suicide over the affair."*

The next chapter provides a brief history of the Earth's climate that provides some context and hopefully some uncertainty when discussing current climate conditions.

Chapter 2 Summary: **The Intergovernmental Panel on Climate Change (IPCC) is the third leg of the global warming cartel (in addition to politicians and media).** The IPCC violates every scientific principle: it treats its hypothesis as a conclusion (*global warming is the result of human-caused CO_2*); it violates the peer review process by using ideological nepotism among reviewers; it alters Working Group scientific results without consulting the originating scientists, and; it presents questionable results as facts, e.g., the hockey stick ruse. And it is silent (remember Gore's lecture on remaining silent?) on those who violate scientific processes and basic ethics including those involved in the Climategate scandal. The content of the leaked Climategate emails should be sufficient evidence for rational people to suspect the motives of the IPCC and its advocates, as it was for Judith Curry and Klaus-Ecken Puls.

The evidence presented in this chapter demonstrates that the IPCC is a bureaucratic entity whose existence depends on continuing CO_2 alarmism without which it has no *raison d'être*, and that advocates of the IPCC and its *cause*, including otherwise reputable U.S. government agencies, will do just about anything to advance the agenda.

The Chapter demonstrates the degree to which AGW alarmists, government agencies and individuals will go to support their ideology, deceiving the public by changing or omitting data, and even misrepresenting scientific facts; for them, *the end justifies the means*.

Chapter 3
Climate History

"The Ice Age Cometh?" This was the March 1, 1975 cover of *Science News*. It shows rampaging glacial ice mowing down skyscrapers, bridges, etc., an apocalyptic scene worthy of a Gore AIT sequel.

From the mid-1940s through the mid-to-late 1970s the Earth was cooling so much that climate scientists, Science News, Time, Newsweek, and other media warned of a pending ice age. Nearly 30 years later, Time unashamedly printed equally disturbing warnings of global warming.

The Earth has experienced climate change in the past without help from humanity. Scientists know about distant past climates because of evidence (proxy data) left in tree rings, layers of ice in glaciers, ocean sediments, coral reefs, layers of sedimentary rocks, etc. For example, ice core samples of the Earth's atmosphere provide a history of temperature and CO_2 variations that stretch back about 800,000 years before present. This chapter provides information on the Earth's historical temperature and CO_2 (Sections 3.1 and 3.2, respectively); the history of temperature and CO_2 measurements (Section 3.3), followed by a presentation addressing the past 100 years of hurricane activity (Section 3.4).

The reason for including a section on hurricanes is that during the 2017 hurricane season, while I was working on this book, hurricane Harvey, the first Category-4 hurricane to strike the U.S. in 13 years, was a slow moving hurricane that caused substantial damage and loss of life. It was, therefore, fertile ground for alarmists to connect it to anthropogenic CO_2, so I researched the subject and my findings are presented in section 3.4 of this chapter.

3.1 Temperature Variations: The following sections present temperature data from 800,000 and 150,000 years ago (YA); 17,000 YA, 10,000 YA; 1,000 YA, and; for the 20th and 21st centuries. The vertical (temperature) scales on graphs labeled Temperature Variation or Anomaly are relative to the average global temperature for the 30-year period 1961 to 1990, which is the "0" level.

800,000 and 150,000 YA to Present: Figures 3-1 and 3-2 show temperature variations from 800,000 and 150,000 years ago to the present, respectively. It is clear from Figure 3-1 that over the past 800,000 years the Earth's *normal* climate is a series of glacial periods that last for about 100,000 years, interspersed by shorter interglacial periods of around 20,000 to 30,000 years; we are living in an interglacial period. In other words, over its hundreds of thousands of year's history the Earth is more naturally a cold planet with relatively short warm periods. The 100,000-year periodicity has been attributed to the Milankovitch Cycles (position and tilt of the Earth relative to the Sun) discussed in Chapter 4, Section 4.1.3.

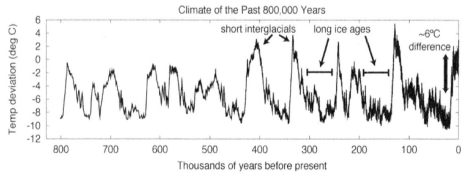

Figure 3-1 Temperatures over the past 800,000 years (Source: Adapted from NASA by Robert Simmon, based on data from Jouzel et al, 2007).

Even the *interglacial* periods, prior to about 400,000 years ago, didn't reach the nominal zero point while glacial temperatures hovered between -8 to -10°C lower than the 30-year average. From 400,000 years ago, however, interglacial temperatures often exceeded current and recent levels by as much as 5.5°C and lasted for several thousand years. The most striking feature of these records is the magnitude of the temperature swings, sometimes as much as 13 to 14°C.

Both figures show that the warmest period was during the last interglacial, around 130,000 years ago, when it was about 5.5°C warmer than the 30-year average. Figure 3-2, an expanded view of the past 150,000 years, shows a cooling trend that lasted around 110,000 years until about 17,000 years ago at which time we entered the Holocene epoch. The Holocene is the name given to the current interglacial; it is the period during which civilization as we know it emerged.

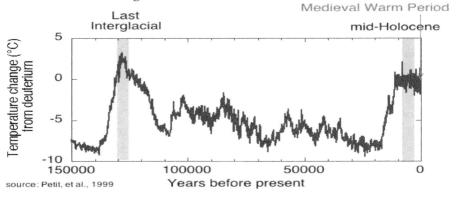

Figure 3-2 Temperatures over the last 150,000 years

(Source: ncdc.noaa.gov/paleo/globalwarming/paleobefore)

<u>17,000 YA to Present:</u> Figure 3-3 shows the wildly varying temperature during the transition from the last glacial to the Holocene. The most abrupt changes occurred in a period known as the *Younger Dryas*, a time of extreme turbulence during which, partway through the transition from the last glacial to the warmer

interglacial, temperatures suddenly returned to near-glacial conditions. The end of the Younger Dryas, about 11,000 to 12,000 years ago, was particularly abrupt. In Greenland (where data were recorded), temperatures rose about 10°C over a few decades. An approximation of the relative current period temperature is indicated by the horizontal bar, marked "Present Temperature," and at the extreme right side of the graph the level of Present Global Warming is also indicated. Temperature variations over the Holocene epoch show clearly that our current 0.8°C warming over a hundred years or so is neither unique nor, should it be, alarming.

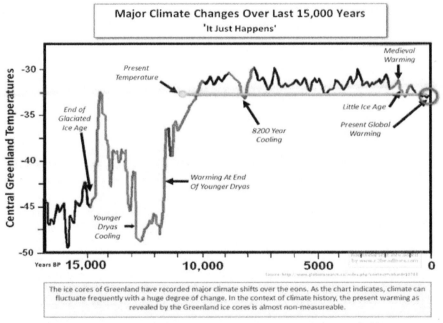

The ice cores of Greenland have recorded major climate shifts over the eons. As the chart indicates, climate can fluctuate frequently with a huge degree of change. In the context of climate history, the present warming as revealed by the Greenland ice cores is almost non-measureable.

Figure 3-3 Temperature Variation over the Holocene (Source: public domain)

The Younger Dryas' return to a cold, glacial climate was first considered to be a regional event restricted to Europe, but according to Professor Emeritus Don J. Easterbrook (*Dept of*

Geology, Western Washington University) et al, later studies have shown that it was a world-wide event. A possible explanation for these extreme temperature oscillations was a change in oceanic circulation, discussed in Chapter 4, Section 4-3.

<u>10,000 YA to Present:</u> Figure 3-4 is a more detailed view of the past 10,000 years. It was generated by geologist David Lappi from data provided by the U.S. government's Greenland Ice Core Project (GISP2). As part of GISP2, the U.S. government drilled the ice core in central Greenland over a five-year period. The data collected reports temperatures every 10 to 60 years which, considering such measurements are often based on millennia increments, is high resolution.

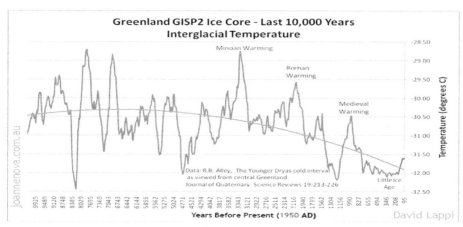

Figure 3-4 Greenland Temperature over the Past 10,000 Years

Figure 3-4 shows three interesting features: (1) there were multiple warm periods prior to the 20th century, including Minoan, Roman, and Medieval, when there was no measurable fossil-fuel related CO_2 in the atmosphere, (2) the historical, relatively low temperature that occurred about 100 years ago (right side of graph, exiting the Little Ice Age (LIA) and; (3) there is an overall cooling trend indicated by the smooth curved line, which was the primary point of Lappi's research, most significantly from the Minoan Warm Period (MWP) to the LIA when the temperature dropped by about 2.8°C. **The 20th century was the first century following the LIA so it is not surprising that, starting from such a baseline, it warmed.**

The curved trend line shows that the Northern Hemisphere (NH) has been experiencing a declining temperatures *trend* over the past 6,000 to 7,000 years or so and we are likely to be heading down toward the next glacial which historically, would be due in the near (geological time) future. The LIA was one of the longest sustained cold periods during the current interglacial, lasting for about four hundred years. **We are recovering from this abnormal cold period, and the recovery began long before anthropogenic greenhouse gases were produced in any quantity.**

Lappi demonstrated a similar pattern in the Antarctic (Southern Hemisphere) over a similar period, again showing the Earth's cooling trend, and it also shows that temperature *hot spots* were not unique to the NH, at times spiking to as high as 2°C above recent values. And though the individual temperature spikes and dips are different than they were in the NH, the long-term temperature trend for the planet appears to be down, not up, which would be consistent with historical glacial/interglacial periodicity.

1,000 years BP: In relatively recent history, warm and cold phases alternated approximately according to thousand-year cycles as shown in Figure 3-4; the Minoan Warm Period three thousand years ago, and the Roman Warm Period two thousand years ago. During the Medieval Warm Period around a thousand years ago, Greenland was colonized (by the Vikings) and grapes suitable for winemaking were cultivated in England (fortunately it cooled and the UK switched to beer!). Cold periods prevailed between warm phases, among them the Little Ice Age, which lasted from the 15th to the 19th century. All these temperature fluctuations occurred at times when only natural atmospheric CO_2 was present. And over the past 1,000 years, temperatures continued to oscillate around a nominal value.

According to Büntgen et al (*American Meteorological Society,* 2017), proxy evidence from the Pyrenees reveals significant Western Mediterranean climate variability since medieval times. They presented the following illustration, Figure 3-5 (Northern Hemisphere Temperatures), which shows temperature variations (relative to June, July, August average from 1961 to 1990) over the past 1,000 years. The trend lines (adapted from the original Buntgen et al graph) are "eye-balled" (I didn't have the data to calculate them) but are close enough to show that 20th century warming is not a new phenomenon and that temperatures repeatedly trend up, down, and steady regardless of CO_2 levels.

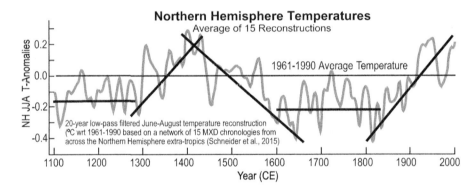

Figure 3-5 Northern Hemisphere Climate Variability since Medieval times

This pattern repeats on shorter time-scales as discussed below.

20th and 21st centuries: Temperatures in more recent times have also varied significantly. Data from National Centers for Environmental Information (NCEI), National Oceanic and Atmospheric Administration (NOAA), ncdc.noaa.gov/cag, hardly a skeptics site, shows the U.S. lower-48-state temperature history as (1) no change from 1895 to 1910, (2) an increase of about 0.8°C from 1910 to 1940, (3) a fall of about 0.4°C from 1940 to late 1970s, and (4) a rise of about 0.9°C from 1980 to 2012, the last date of record.

Given the nearly 1°C increase from 1980 to 2012 it is not surprising that in the absence of historical data and perspective many were alarmed, and encouraged to be alarmed by the establishment of the IPCC at the peak of this warming trend (1988) with its apocalyptic messaging.

From 2000 to 2011, according to Hadley Centre in the UK (ocean data) and the Climate Research Unit (CRU) of the University of East Anglia (land-based temperatures), and Remote Sensing System (satellite data), the temperature trend was slightly negative, i.e., the average global temperature cooled over this 11-year period. Satellite data gathered by the University of Alabama, Huntsville, showed a slight temperature increase over that same period. None of the trends was statistically significant, i.e., no temperature change occurred over that period. A follow-up analysis by RSS covering an extended period from February 1997 through November 2015, also based on satellite data, showed no global warming increase. Of course, various regions did experience warming (and cooling), but global average temperatures were constant during this period; while manmade CO_2 was increasing, year over year. If the AGW hypothesis was correct shouldn't the temperature increase year over year?

Figure 3-6, provided by the CRU, shows the almost steady-state (average) temperature from August 1997 through October 2011. Note, the steady-state temperature is 0.42°C above the nominal 30-year average. It also shows the recorded temperature oscillations over the period.

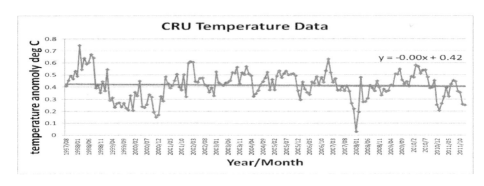

Figure 3-6 CRU Temperature Data from 1997 through 2011

While the average global temperature over this period was stable, it is clear from Figure 3-6, that temperatures oscillated above and below the steady-state line, sometimes quite significantly, while CO_2 steadily increased over the past 60 years, as shown in Figure 3-7 (section 3.2). These data do not bode well for those who believe the relationship between anthropogenic CO_2 and temperature is dominant; clearly, it is not.

Many of the extreme high and low variations have been attributed to El Niño/La Niña events (natural atmospheric and oceanic oscillations discussed in Chapter 4, Section 4.4). Since 1952, NOAA has classified the severity of El Niño into four categories, *weak, moderate, strong,* and *very strong,* and La Niña into three categories, *weak, moderate,* and *strong.*

Three *very strong* El Niño occurred in 1982/83, 1997/98, and 2015/16, each of which did coincide with significant warming. Only the 1997/98 El Niño temperature spike (extreme left of graph) appears within the timescale of the Figure 3-6 graph; the severity of that El Niño has been described as *"the climate event of the twentieth century."*

On the other side of the equation, *strong* La Niña did occur in 2007/08 coincident with the 0.4°C drop shown on the graph, and the 2010/11 decrease of about 0.2°C was also coincident with the *strong* 2010/11 La Niña.

While it is generally accepted that globally the average temperature has increased in recent years, regionally, many countries had different experiences. In 2008, Saudi Arabia and the deserts of Iran and Iraq experienced snow and the coldest temperatures in recent memory of -5°C in Riyadh, 16°C degrees colder than the average. Great Britain experienced the coldest winter in 30 years, the winter of 2009-10, with the British media calling it *"The Big Freeze."* And the winter of 2011-12 set a large number of cold records; the U.K. and many Swedish cities recorded the coldest December since record-keeping began. In Northern Ireland the temperature plunged to -18°C, a record low. Germany had its coldest December in 40 years. In February 2012, a deadly cold wave swept over eastern and northern Europe where temperatures as low as -39°C were recorded and, in 2012, Russia had its coldest December ever, with the temperature in Siberia dropping to -60°C. All these 20th and 21st century variations occurred during times when human-induced CO_2 was steadily increasing.

Among recent media hysteria in the UK about this year's (2018) heat wave, are arguments that not enough is being done to combat carbon dioxide increases. And then, the media say the 2018 heat wave is the worst *since* 1976! In 1976, CO_2 was about 350 ppm, considered by Gore, Hansen and other alarmists to be an ideal level, yet there were extreme heat waves throughout Europe. And, according to Wikipedia, in 1911, a heat wave hit the UK with temperatures around 36°C (97°F). The heat began in early July and didn't let up until mid-September, and even in September temperatures were still up to 33°C (92°F). At that time CO_2 levels would have been around 300 ppm, almost at the pre-industrial level (280 ppm).

3.2 Carbon Dioxide (CO_2): Figure 3-7 from the NOAA Earth System Research Laboratory (Jan 2017), shows a steady linear increase in atmospheric CO_2, from 1958 to 2017 of about 90 ppm over 60 years, i.e., about 1.5 ppm per year average. At that rate it will take more than 200 years for CO_2 to double its 280 ppm reference value or another 100 years from the present level (400 ppm). Furthermore, as discussed above, over the 60-year period temperatures have risen and fallen (from the 1970s next Ice Age fears to the warming of the 1990s).

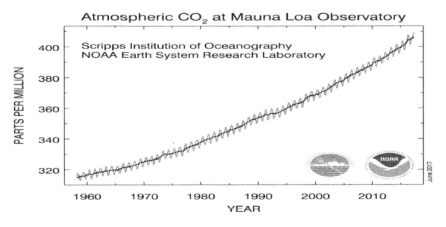

Figure 3-7 CO_2 Variation over the past 60 years (Source: NOAA)

But what does that matter if the CO_2 argument is just a dissemblance for another agenda, as illuminated by Richard Benedick, U.S. chief negotiator for 1987 Montreal Protocol on Ozone-Depleting Substances and Special Advisor to Secretary-General (Maurice Strong) of the United Nations Conference on Environment and Development (Rio de Janeiro, 1992), said,

"A global warming treaty must be implemented even if there is no scientific evidence to back the greenhouse effect." The end justifies the means!

According to geologist Professor G. Robinson et al (2009, *Global Warming: Alarmists, Skeptics and Deniers; A Geoscientist Looks at the Science of Climate Change*), paleo-climatologists have determined from ice core samples that the current concentration of CO_2 is the highest level for the length of time ice core records represent, approximately 800,000 years which, at first glance may cause pause for skeptics; however, read on.

Ice core samples from 800,000 years ago to recent times show a steady increase in CO_2 concentrations: 175 parts per million (ppm) about 670,000 years ago (note, according to Greenpeace cofounder Dr. Patrick Moore, at 150 ppm CO_2 all plants would die, resulting in the end of life on Earth); 300 ppm 320,000 years ago, and today at about 400 ppm; i.e., CO_2 has more than doubled over that period.

The increase from 175 ppm to 300 ppm occurred before there were humans (Homo sapiens). The Industrial Age began about 150 years ago, and we've only had automobiles for about 100 years. An interesting couple of questions might be (1) what caused the steady increase of CO_2 over the 800,000-year period? And, (2) **why do alarmists seem to believe that nature stopped contributing CO_2 to the atmosphere at the beginning of the Industrial Age?**

From a geological perspective, Dr. Robinson states the ice core records are relatively *recent* data and don't tell the whole story. He says that the amount of CO_2 in the atmosphere today is *abnormally low* when considered from the viewpoint of Earth history and *geologic time*.

Citing a study by Berner and Kothaval (2002, *Geocarb III: A revised model of atmospheric CO_2 over Phanerozoic Time; American Journal of Science*, Vol. 301), Robinson says that during the Paleozoic Era, from about 550 to 350 million years ago (mya), while life was restricted to the oceans, CO_2 fluctuated from 10 to 25 times higher than today's levels. It dropped sharply in response to the appearance and spread of land plants about 350 mya, then shot up again during the Mesozoic Era, the Age of the Dinosaurs. From a peak value of more than ten times the current levels during the Mesozoic, CO_2 began a long, gradual and irregular decline to the low levels seen in the ice core samples from around 800,000 years ago.

Figure 3-8 shows CO_2 variation over the last 500+ million years, from multiple sources (identified within the graph). The curves show remarkable consistency in depicting much higher CO_2 levels historically than over the past 150 years or so. The graph begins (on the right) with an era predating terrestrial plant life, during which *solar output* was approximately 4 to 5% lower than today. (According to *Universe Today*, The Sun increases its luminosity by about 1% every 100 million years - it is currently at about 30% of its final level). Toward the left side of the graph the Sun gradually approaches modern levels of solar output, while vegetation spreads, removing large amounts of CO_2 from the atmosphere (photosynthesis). At the far left of the graph, we see modern CO_2 levels and the appearance of the climate under which human species and human civilization developed. At this point, we might be wandering if perhaps the Sun has something to do with climate change (recall from the Chapter 2 NASA article "What Are The Primary Forcings of The Earth System?" - *"The Sun is the primary forcing of the Earth's climate system."*).

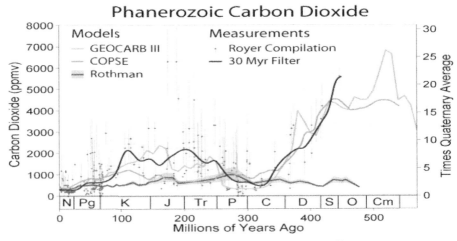

Figure 3-8 CO_2 Variation over the past 500 Million years

(Source: public domain)

Figure 3-9, is a reconstruction of temperature and CO_2 over the past 10,000 years from ice cores, using data from Alley (2000, *The Younger Dryas cold interval as viewed from central Greenland, Quaternary Science Reviews* 19, 213-226).

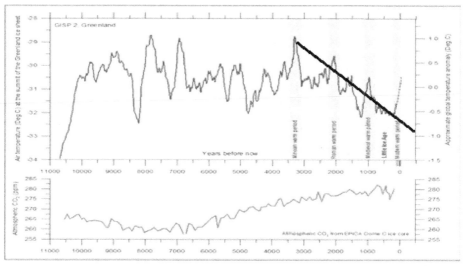

Figure 3-9 CO_2 versus Temperature over the past 10,000 years

(Source: public domain)

The top panel of the diagram shows that Greenland (Northern Hemisphere) experienced numerous cycles of warming and cooling over the past 10,000 years that involved temperature variations of about 5.5°C. As with Lappi's Figure 3-4, it shows a fairly significant declining temperature trend beginning a little more than 3,000 years ago from the Minoan Warm Period (MWP) that culminated in the LIA (I added the trend line from the MPW to the LIA). The dotted line (extreme right-hand side) provides an approximate context to the late 20th-century warming. Again, not only does it show that there is nothing unusual about the current episode of increased global temperature in either timing or amplitude, it reveals a necessary (to avoid a glacial episode) reversal from the cooling trend following the MWP.

The bottom panel shows CO_2 levels steadily increasing over the same period, again, indicating no apparent causal correlation between the two records.

Evidence that CO_2 Lags Temperature Increases: Figure 3-10 shows the relationship between CO_2 and global temperature for the past 400,000 years. **This graph has been used by advocates of the human-caused global warming argument to mislead the public** as discussed in Chapter 2 under the heading Articles of Deception, including by Al Gore in his AIT movie, and the mislabeled version in David and Gordon's children's book. The record, reading from right to left, shows that over hundreds of thousand-year periods CO_2 (lighter line) does not lead temperature but rather, lags temperature (darker line).

One explanation is that far more CO_2 is dissolved (and stored) in the oceans (comprising about 70% of the Earth's surface) than in the atmosphere, and oceans retain CO_2 over much longer periods. Because of their volume, and depth, it takes much longer for oceans to warm from increasing temperatures than for land. Colder (denser) water absorbs more CO_2 than warm water. Then, as temperatures increase, warming oceans slowly release more CO_2 into the atmosphere. This process is discussed in Chapter 4, Section 4.6, Carbon Cycle.

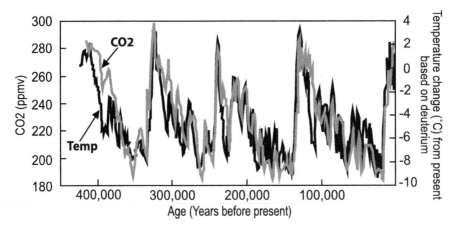

Figure 3-10 CO_2 vs Temperature over the past 400,000 years

(Source: Adapted from U.S. Dept of Energy)

3.3 Temperature and CO_2 Measurement History: Throughout this book, I have accepted that temperature increased over the 130 years or so by a nominal 0.6 to 0.8°C, and that CO_2 increased over that timeframe from 280 ppm to around 400 ppm. These are the figures that both sides of the debate generally use and so my attempt to create *uncertainty* about knowing the cause of climate change was based on these values. However, curiosity drove me to research the basis for the numbers and my conclusions only add to a case for uncertainty. While it is clear that both the temperature and CO_2 levels have increased from where they were at the end of the 20th century, the precise amount of increase is not clear. Following is what I found:

Temperature: There are three methods for measuring temperature: thermometers, radiosonde balloons, and satellites. Thermometers measure surface temperatures at thousands of locations around the world but mostly concentrated in the industrialized nations of the USA, Europe, and S.E. Asia. According to the World Meteorological Organization (WMO), in 2017 there were about 11,000 thermometers housed in weather stations around the world (which has a surface area of approximately 195 million square miles). Furthermore, weather stations are on land (29% of the Earth's surface area); there are no stations on the surface of the ocean (71% of the Earth's surface), which until the 1940s was measured from buckets of water by liquid (mercury or alcohol) thermometers.

Radiosonde, which measures the atmosphere (temperature, pressure, humidity) at different altitudes, has been operating since the 1950s and according to the WMO there are about 1,300 stations globally. Satellites have monitored global temperatures since 1979; they cover the entire globe. But since global temperature data from land-based thermometer measurements from the mid-1800s are used to determine the long-term temperature trend, this discussion focuses on the thermometer aspect of data gathering in which there are many potential sources for error, including:

1) Calibration – thermometers graduated in degrees are at best accurate to +/- 0.5°C; this alone could account for most the warming. Furthermore, the various thermometers are not calibrated to a universal standard.

2) Many climate stations do not meet the requirements of the World Meteorological Association (WMA) requirements for correct *siting*. Pielke et al (2007, *Documentation of uncertainties and biases associated with surface temperature measurement sites for climate change assessment, Bulletin of the American Meteorological Society*) found that in the USA, **50% of the stations are incorrectly sited**, saying,

"...there are large uncertainties associated with the surface temperature trends from the poorly sited stations.... the use of the data from poorly sited stations provides a false sense of confidence in the robustness of the surface temperature trend assessments." Imagine the siting issues with measuring stations in most other countries.

And according to Wikipedia, temperatures on a warm summer day can vary by 16 to 17°C (30°F) from just above the ground to waist height. This begs the question, are all stations located at the same height? Probably not.

3) There is a high density of weather stations in the USA, Europe, and S.E. Asia and vast gaps in Africa, South America, the Middle East, India, Greenland, Siberia, the Arctic tundra, and Antarctica.

4) Measurement issues associated with the concentration of weather stations in developed countries is exacerbated because many stations that were once in rural areas, because of population increases, became urban areas and subject to the *urban heat island* (UHI) effect. UHI is the increase in temperature due to increased amounts of concrete, asphalt, buildings, pavements, air conditioners, heating, autos, and other city infrastructure all of which increase with increasing population and which change the amount of the Sun's energy reflected back into space (albedo) or absorbed by the Earth. According to Pielke et al (2002, *The influence of land-use change and landscape dynamics on the climate system*), in regions of intensive human-caused, land-use change such as North America, Europe and southeast Asia, the

"local change in albedo may actually have a greater effect on the energy budget than that due to all the well-mixed anthropogenic greenhouse gases together."

Chapter 4, section 4.1 demonstrates the sensitivity of global temperature to albedo – a 3% variation in albedo can have a 3°C impact. And, according to Spencer (2017, *Climate Change: The Facts*), no one has yet found a way to remove the very local UHI warming effect from thermometer and thermostat data.

5) There have been enormous changes in the number of weather stations over the years causing data disruptions.

According to Plimer (2009, *Heaven and Earth: Global Warming the Missing Science*), when the Soviet Union collapsed and weather stations were shut down, the USA showed a sudden increase in temperature. The Soviet Union had many *rural* stations.

6) The surface temperatures are taken from grid boxes varying from 120 x 120 (14,400) square miles to 300 x 300 (90,000) square miles. NASA Goddard Institute for Space Studies (GISS) assumes stable temperatures within a 750-mile radius of a weather station. We see significant temperature variations between Denver and Colorado Springs, which are about 70 miles apart. For instance today, 25 July 2018, there is a 12°F (about 6.7°C) difference in temperature, Denver registering 74°F and Colorado Springs at 62°F.

7) Time of day that recordings were made – if the times were not consistent then the data were invalid, and there is no possible way to know for certain when temperatures were recorded. This is important because diurnal variations can be significant and, in addition to the position of the Sun, humidity varies during a day. According to Wikipedia, high desert regions typically have the greatest diurnal-temperature variations, while low-lying humid areas typically have the least. For example, the Snake River Plain, in Idaho, can have high temperatures of 38 °C (100 °F) during a summer day, and then have lows of 5 to10 °C (41to 50 °F). At the same time, Washington D.C., which is much more humid, has temperature variations of only 8°C (14 °F).

8) Weather stations are not located randomly, a requirement for statistically reliable data, the other constraints notwithstanding.

All the above, which is not an exhaustive list, can cause temperature data to be contaminated. According to Ross R. McKitrick and Patrick J. Michaels (2007, *Journal of Geophysical Research-Atmospheres*), **scientists readily acknowledge that temperature measurements are contaminated**, but they, the scientists, argue that there are adjustments to fix the problem. For example, to deal with a false warming generated by urbanization, the UHI effect, they have the *Urbanization Adjustment*. To deal with biases due to changing the time of day when temperatures are observed, they have the *Time of Observation Bias Adjustment*. To deal with the loss of sampling coverage they have the *Coverage Adjustment*. And so on. Again, as with climate models the debate is essentially adjudged by statisticians and modelers. Also, as McKitrick and Michaels point out, **each time historical data are analyzed and *adjusted*, it gets warmer!** Recall that all 35 *errors* in Gore's AIT were in the same direction.

How do we know these adjustments are correct or adequate? According to McKitrick and Michaels, the reasoning by AGW advocates is, in effect, we do not need to worry about urbanization bias for instance because experts applied the Urbanization Adjustment – i.e., *circular logic*. A few studies argue that the adjustments must be adequate since adjacent rural and urban samples give similar results. But Spencer (2017) shows that station warming bias is very sensitive to even small population increases. For instance, he shows a 0.6°C increase in UHI effect between a completely rural (people free) environment to one with a population of only 1,000 per km^2 (about 390 square miles) and 0.6°C is the lower end of the range that accounts for the entire warming over the past century or so.

Conclusions about the amount of global warming, and the role of greenhouse gases, are based on the assumption that the *adjustment models* work perfectly. Scientists who attribute warming to greenhouse gases argue that their climate models cannot reproduce the surface trends from natural variability alone. And, if they can't attribute it to natural variability, they attribute it to greenhouse gases, since (they assume) all other human influences have been removed from the data by the adjustment models. This is of course a *logical fallacy*.

And if the above is insufficient to create uncertainty, how about a NASA GISS FAQ website which says that the most trusted models produce a global average temperature of roughly 14°C (57.2°F), but that it may easily be anywhere between 56 and 58°F. The difference between 56°F and 57.2°F is 1.2°F or 0.7°C, almost the entire warming increase over the past hundred years or so (data.giss.nasa.gov/gistemp/faq/abs_temp).

Carbon Dioxide (CO_2): In 1959, the measurement of atmospheric CO_2 changed from one based on chemical analysis, called the *Pettenkofer Method*, which spanned about 180 years of collection, to one based on infra-red sensor analysis (spectroscopy), the latter with its primary measuring station located at the Mauna Loa facility in Hawaii.

There are many articles that criticize Mauna Loa data based on it being located on an active volcano and that its data were not validated against the 180 years or so of Pettenkofer data. Most of these articles suggest that there has not been a steady increase in CO_2 since the industrial revolution, but that it has varied significantly and in such a manner as to question its relationship to human emissions. For instance, according to Beck and Freiburg (2006, *180 Years accurate CO_2 – Gas Analysis of Air by Chemical Methods*), since 1812, CO_2 concentration in the Northern Hemispheric fluctuated from 400 ppm in 1829, to 300 ppm in 1900, to above 400 ppm in 1942, the latter value about the same amount as today measured from Mauna Loa. Beck et al claim accuracy for chemical analysis of around 3%.

An article by Dr. Timothy Ball (2008, *Measurement of Pre-Industrial CO_2 Levels*) asserts that both historical and Mauna Loa CO_2 data are *drastically* modified to show pre-industrial levels were approximately 270 ppm. Primer (2009) says that the IPCC pre-industrial start value of 270 ppm (I have also seen 280 ppm) is based on G.S. Callender's research (1938, *The Artificial Production of Carbon Dioxide and its Influence on Temperature: Quarterly Journal* *of the Royal Meteorological Society*) that derived the number from chemical methods, no longer accepted by the IPCC. As Plimer says,

"They (the IPCC) are prepared to use the lowest determination by the Pettenkofer method as a yardstick yet do not acknowledge the Pettenkofer method measurements showing CO_2 concentrations far higher than now many times since 1812." I believe this is called "cherry picking."

Dr. Zbigniew Jaworoski, MD, PhD, DSc, a multidiscipline scientist (1997, *Another Global Warming Fraud Exposed: Ice Core Data Show No Carbon Dioxide Increase*) says,

"from the very beginning, the hypothesis on anthropogenic greenhouse warming was tainted with a biased selection of data, ad hoc assumptions that were not verified experimentally, and one-sided interpretations."

He also attributes the source of the IPCC pre-industrial CO_2 level of 270 ppm to research by Callender who, in his article presented to the Royal Meteorological Society (RMS) claimed that because of fossil fuel burning, the average atmospheric concentration of CO_2 had increased from the 19th century value of 274 (+/-0.05) ppm to 325 ppm in 1935.

However, Jaworoski says the measured 19th century CO_2 concentrations in the atmosphere ranged from 250 to 550 ppm and that the average concentration estimate was 335 ppm and, that to reach the low 19th century CO_2 concentration, which became the baseline for the global warming hypothesis, it appears that **Callender purposefully selected data that supported his hypothesis** and omitted large amounts of significant probative data. From 90,000 samples and a set of twenty-six 19th century averages, he says, Callender rejected 16 that were higher than his assumed low global average and two that were lower. Figure 3-11 shows the values Callender chose to use (circled) and those that were omitted. It is clear from this graph that he omitted a large amount of relevant data that if included could have changed the entire anthropogenic CO_2 debate.

Callender's paper was criticized by several Royal Meteorological Society members who challenged many fundamental aspects of it including the validity of the estimated CO_2 concentration, and commented that Callender had not demonstrated that changing one factor (CO_2) could be responsible for changing temperature. The above analysis of temperature and CO_2 baselines, from which projections are made, should create further uncertainty in the reader.

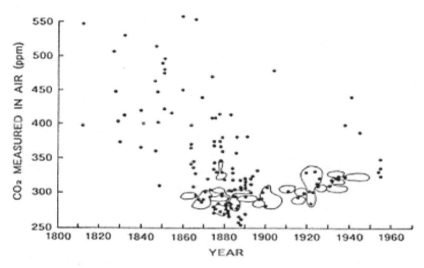

Figure 3-11 Average 19th and 20th Century Atmospheric CO_2 Concentrations

Short term extreme weather events neither confirm nor refute the accuracy of climate prognoses. Only a systematic and worldwide evaluation, spanning multiple decades, hundreds, and even thousands of years, can reveal meaningful trends that isolate data from statistical noise. This is evident from the above and also when considering the recent hurricane activity in the Atlantic.

3.4 Hurricanes: Within two weeks of hurricane Harvey striking land, August 20th 2017, Nicholas Kristof of the New York Times, Sep 2, 2017, wrote: *"We can't have an intelligent conversation about Harvey without also discussing climate change."*   When global warming alarmists introduce the subject of climate change, they typically mean "human-caused CO_2 emissions" since, as repeated several times, skeptics in the main do not contest that the Earth has been in a warming trend from the mid-19th century, only that the causes are mostly part of natural cycles. In any case, alarmists attribute any significant weather event to be the *human-caused*. Harvey, followed by Irma and Maria were devastating on the communities struck by those hurricanes. However, for an *"intelligent"* conversation to occur we need a baseline against which recent activities can be compared; i.e., a historical perspective.

Since neither Mr. Kristof nor I know enough about the science (Wikipedia says he's a journalist; who studied government at Harvard and received a Law degree from Oxford), I refer to data from the Hurricane Research Division, Atlantic Oceanographic and Meteorological Laboratory, National Oceanic and Atmospheric Administration (NOAA), and other meteorological experts.

According to a research paper by Gabriel A. Vecchi and Thomas R. Knutson (December 2007, *NOAA/Geophysical Fluid Dynamics Laboratory,* Princeton, New Jersey):

"The trend in average Atlantic Tropical Cyclones duration over the period from 1878 to 2006 is negative … even though Sea Surface Temperature has warmed significantly."

Note, this was published just two years after a very active 2005 hurricane season. In fact, the 2005 Atlantic hurricane season was the most active Atlantic hurricane season in recorded history, shattering numerous records (when CO_2 levels were about 20 ppm lower). Of the storms that made landfall that year, four of them were Category-3 hurricanes, Dennis, Katrina, Rita, and Wilma, The impact of the hurricanes was widespread and ruinous, with almost 4,000 deaths and damage of about $160 billion, which is almost certainly why Gore's movie of 2006 had such traction; they were horrific and at the time, recent.

Table 3-1 was constructed from data provided by the National Hurricane Center (NHC), NOAA (aom1.noaa.hrd.tcfaq). It shows the number of major hurricanes (Categories 3, 4 and 5) that struck the U.S. by category, by decade for the 20th century, the decade from 2000 to 2009, and the next seven and a half years. Over the hundred years of the 20th century the decadal average was 6.4. At the time of writing this section, with about 2.5 years left in the decade, the average for this century is 5.0.

While these data reflect only the major hurricanes that made landfall, according to NOAA, prior to the implementation of weather satellites they represent a more reliable account of events than do the numbers for total hurricanes in the Atlantic prior to satellite data. The same (NOAA) data source shows a similar pattern for all hurricanes, categories 1 through 5, regardless of whether they made landfall.

It is interesting to note that the prevailing global average climate conditions, warm or cool, do not appear to have an effect on the annual number of hurricanes. During *cool* decades, the number of major hurricanes was slightly above average (6.6), while so far the average for the *warm* periods is 6.0. In other words, no statistical difference. And the same conclusion for CO_2 – while the average number of major hurricanes stayed relatively constant, CO_2 increased from about 280 ppm to 400 ppm.

Table 3-1 Major Hurricanes that Struck the U.S by decade

Decade	Cat 3	Cat 4	Cat 5	Total Major	Climate
2010 - 2017	0	1	1	3	Warm

2000 - 2009	6	1	0	7	Warm
1990 - 1999	4	0	1 (1992)	5	Warm
1980 - 1989	5	1	0	6	Warm
1970 - 1979	4	0	0	4	Cool
1960 - 1969	3	2	1 (1969)	6	Cool
1950 - 1959	7	2	0	9	Cool
1940 - 1949	7	1	0	8	Cool
1930 - 1939	6	1	1 (1935)	8	Warm
1920 - 1929	3	2	0	5	Warm
1910 - 1919	5	3	0	8	Warm
1900 - 1909	4	1	0	5	Cool

Dr. Roy Spencer, whose PhD is in Meteorology (his dissertation was on the structure and energetics of incipient tropical cyclones) and who developed a method for monitoring the strength of tropical cyclones from satellites, examined all of the major hurricane strikes in Texas since 1870 and plotted them on the time series of Sea Surface Temperature (SST) variations over the western Gulf of Mexico (Figure 3-12). As can be seen from the graph, the occurrence of major hurricanes is indifferent to whether the Gulf is above or below average temperature.

His explanation is that hurricane formation requires a unique and complex set of circumstances to occur, and sufficiently warm SST is only one of them.

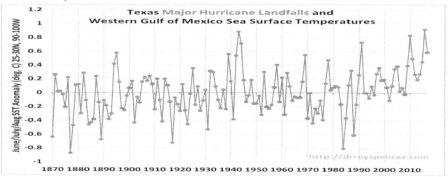

Figure 3-12 Major Hurricanes Strikes in Texas versus Sea Surface Temperatures

This explanation is consistent with one by Gerry Bell, Ph.D., Lead Seasonal Hurricane Forecaster with NOAA's Climate Prediction Center who predicted 2 to 4 major Atlantic hurricanes in 2017 because he said, El Niño was weak, and a strong El Niño together with strong Wind Shear - Wind shear refers to a change in wind speed or direction with height in the atmosphere - typically suppress development of Atlantic hurricanes. Remember, there was strong El Niño in 2015/16, two of the hottest years on record, when no major hurricanes occurred.

According to all the referenced sources, **scientists do not know what causes hurricanes to form.** They do know, and agree, that there are two minimal conditions that need to be present, hot humid air and wind. The Sun warms the ocean around the equator and when the temperature reaches about 80°F (27°C) it evaporates into water vapor, rises, cools with altitude, forms clouds, which are carried by wind; in the case of Atlantic hurricanes, by Trade Winds. Once hurricanes are formed and traverse across the ocean, warm water fuels the *intensity* of hurricanes.

To return to Mr. Kristof's call for an *intelligent conversation*, by which I'm confident he means *"the science is settled, it's human-caused CO$_2$, so what are we going to do about it,"* a regression of the NOAA CO$_2$ data, shown earlier in Figure 3-7, would indicate that in 1900, CO$_2$ levels were significantly lower than even the most aggressive IPCC targets, around the pre-industrial benchmark of 280 ppm. And it is generally agreed by both sides of the debate that in 1900 the average global temperature was about 0.8°C cooler than it is today. So what is so special about 1900? Well, according to NOAA, the deadliest natural disaster in American history was the 1900 Galveston Category-4 hurricane which killed an estimated 8,000 people and destroyed more than 3,600 buildings.

Anecdotally then, increased temperature alone does not cause hurricane activity, and clearly higher CO_2 levels do not increase hurricane activity. While many scientists agree that warm SST increases the *intensity* of hurricanes, the NOAA data in Table 3-1, Vecchi and Nutson research (2007), and Dr. Spencer's data in Figure 3-12 don't appear to support that position.

Al Gore, in his AIT movie, didn't actually say global warming caused hurricane Katrina, but he showed the devastation of Katrina in the context of warmer sea surface temperatures, and clearly alluded that it was the result of human-caused CO_2. He dramatically exhibited frightening visuals of Katrina passionately saying,

"There were warnings that this hurricane, days before it hit, would breach the levies and cause the kind of damage that it ultimately did cause."

Yes, but he forgot to mention that his dramatic statement had nothing to do with CO_2 emissions, the target of his movie presentation, it was government ill-preparedness at work; a government for whom he had been vice president for eight years from 1992 to 2000. I'm going out on a limb, but I would guess the levees didn't suddenly deteriorate between 2000 and 2005. He callously attempted to wrap the entire blame for the cause, the effect, and lack of response on human-caused (CO_2) global warming.

He also said that the historical increasing hurricane and flooding *cost* is the *"unmistakable economic impact of global warming"* when even the IPCC (2007) acknowledges that such costs are largely the result of rising population and infrastructure in coastal regions.

Major hurricanes *are* in fact causing more loss of life and increased costs of infrastructure damage, not because of global warming, and not because of increasing CO_2, but because increased populations are occupying more coastal land and building more expensive residences, encouraged by government idiocy.

Since 1968, the U.S. federal government has provided subsidized insurance for homeowners who live in flood-prone areas, a program known as the National Flood Insurance Program (NFIP). According to a 2017 *Politico* article by Ike Brannon and Ari Blask, as of 2016 the NFIP has over 5 million policies in force and subsidizes policyholders to the tune of $3 billion annually. An example of the insanity of such a program was reported by Kelly Swanson in a 2017 article (*The Weeds*), in which she reports: *"there's this one $69,000 home in Mississippi that has flooded 34 times in 32 years and has received $663,000 in payments."* You have to love lobbyists!

Mr. Kristof's response to *Harvey* is a perfect example of both the biases discussed in Chapter1, the *confirmation bias*, and the *recency bias*. Harvey was vicious; it confirmed an *a priori* belief that the (anthropogenic CO_2) warming climate caused it, and of course it was recent when he wrote the article.

Chapter 3 Summary: **There is nothing unique or abnormal about the 20th and 21st century climate.** This chapter demonstrated that today's climate is neither unique nor abnormal when viewed over an appropriate period; in fact, **there is no *normal* temperature or climate for the Earth**. If there was a normal climate, based on 800,000 years of historical records, it would be mostly glacial.

Over the past 800,000 years during interglacial periods, temperatures have exceeded modern day levels and for most of that time there was no human induced CO_2. We are, it appears, living in a relatively benign climate.

Historical records show that CO_2 levels have been significantly higher than they are today, and the science shows that **over the past 400,000 years CO_2 lagged temperature.**

But most importantly, while CO_2 increased linearly over the past hundred years or so, temperatures fluctuated from warnings in the 1970s of a return to the next Ice Age, to warnings in the 1980s through late 1990s that we will warm the planet to extreme and uninhabitable levels in the next century, to the IPCC-acknowledged *pause* in temperature increase; i.e., no increase from about 1998 through 2014.

There is empirical evidence that when viewed over a 7,000 to 8,000 year period, the Earth is on a cooling trend (and clearly we are not preparing for such an eventuality).

The historical records of temperature and CO_2 levels since the beginning of the Industrial Age, on which computer model trends are based, are questionable. Temperature records are contaminated by many factors that are only reconciled by *adjustments* to the raw data – that always support the anthropogenic global warming argument – and for CO_2 measurements, there is a great deal of controversy since 180 years of measurements using a chemical process are discounted, except for cherry-picking to set the pre-industrial baseline.

Despite recent activity, there appears to be no correlation between the temperature of the planet and the *number* of hurricanes over the past century, and **there is no correlation or relationship between CO_2 levels and hurricane activity.**

The cause(s) of hurricanes is still unknown but because of 24/7 media coverage and the devastation of hurricanes on communities, they are fertile opportunities for alarmists to deceive and mislead the public that generally craves certainty.

Chapter 4
The Science

Climate is the typical state of weather at a particular location and time of year, measured over a long period of time; tens, hundreds, and thousands of years. It includes the average of such variables as temperature, humidity, windiness, cloudiness, precipitation, etc. Even major events such as hurricanes, blizzards, tornadoes, floods, etc. are natural climate conditions in certain regions.

The Earth's climate is determined by the flows of energy, originating from the Sun, into and out of the planet and to and from the Earth's surface. When the flow of incoming solar energy is balanced by an equal flow of heat to space, Earth is in radiative equilibrium, and the average global temperature is relatively stable. Anything that increases or decreases the amount of incoming or outgoing energy disturbs Earth's radiative equilibrium and global temperatures rise or fall in response. Such destabilizing influences are called *climate forcings*, including:

- The Sun (4.1)
- Tectonics (4.2)
- Atmospheric and oceanic circulations (4.3)
- Oceanic oscillations (4.4)
- Clouds & Aerosols (4.5)
- Greenhous gases (4.6)

This chapter addresses these forcings in the sections shown in parentheses. Each section concludes with a summary. Section 4.7 provides a macro-view of the Earth System.

4.1 The Sun:

"Is the source of life on Earth. Of the total energy contribution to the Earth's climate, 99.98 percent originates from the Sun." (2015, Vahrenholt and Luning, *The Neglected Sun*).

And 99.85% of the *mass* of our solar system is concentrated in the Sun. The remaining 0.15% is distributed among the planets with Jupiter and Saturn taking the largest share (92%). Consequently, the orbit of the Earth is dominated by the gravitational force of the Sun, although it is influenced by the gravitational effect of the other planets. The Sun's position itself however, is not fixed (DaVinci was wrong!); it also varies with the gravitational pull of other planets, mainly Jupiter and Saturn and it, and the solar system, hurtle through the Galaxy at about 500,000 mph. The Galaxy is also shooting through space at about 1.3 million mph.

The Sun emits energy in the form of radiation, electromagnetic waves extending from low-energy infrared to visible light, ultraviolet, and beyond to x-rays. This energy travels through space, its strength diminishing with growing distance from the Sun as it fans out over an ever-increasing area.

The Earth's atmosphere has several layers ranging from its surface up to about half way to the Moon, but we are only concerned with the two lower layers, the troposphere and stratosphere. The troposphere is the part of atmosphere in which we live and where all weather occurs: it extends from the ground up to about 6 – 10 miles above sea level; the stratosphere extends from the top of the troposphere to about 31 miles above the Earth's surface.

When solar radiation penetrates the stratosphere the ultraviolet component interacts with oxygen to form ozone. The ozone layer provides a protective shield against further UV rays which are mostly filtered out and converted to heat. In the troposphere, certain infrared wavelengths are filtered out by the atmosphere. The remaining infrared (wavelengths) pass through the atmosphere reaching the Earth's surface. Visible light (sunlight) is largely spared the absorption effect of the atmosphere. However, it doesn't all get through to the surface; about 20% is absorbed in the atmosphere and about 30% is reflected back into space by clouds, aerosols, other atmospheric gases, and the surface itself. The remaining radiation, about half the original amount reaches, and is absorbed by, the Earth's surface about 8 minutes after it left the Sun, where is begins a complex series of ecological interactions.

The Sun's energy at the top of the atmosphere is often referred to as the *solar constant*, although it is not a constant; it varies with the Earth's elliptical orbit (changing the distance the Sun's energy travels to reach the Earth) and with the intensity of solar activity. According to IPCC AR5, the solar constant had an average value of around 1,361 Watts per square meter (1360.8 ±0.5) during 2008 although as we will see, as important as this metric is to understanding climate, it is far from an established fact. This measure of energy is called the Total Solar Irradiance (TSI). TSI represents the Sun's effective radiation level after it has travelled the approximately 93 million miles to the top of the Earth's atmosphere (TOA). When dealing with the Earth's radiation budget, the top of the *troposphere* is considered to be the TOA.

The temperature of the Earth without an atmosphere is estimated to be about -18°C, yet the average global temperature at the surface is around 15°C, an increase of 33°C (92°F). This is due to the greenhouse effect particularly that of water vapor and carbon dioxide, discussed in section 4.6.

The amount of sunlight received at the Earth's *surface* is affected by:

- The Sun's surface temperature and its distance from the Earth

• The Earth's shape – results in an average radiation of around 340 W/m² affecting the Earth's surface and atmosphere. (Explanation: if the Earth was a flat, one-sided disc facing the Sun, every square meter of the Earth's surface would receive 1,361 watts of energy from the Sun. However, the Sun's energy is spread over the surface of the Earth which is a sphere with an area four times as great as the area of the disc of the same radius. So the 1,361 W/m² is reduced (divided by four) to 340 W/m² as the average value over the entire surface of the planet, both hemispheres, the four seasons, day and night.

• Orbital variations - there is about 3.5% difference between radiation received when the Earth is at its maximum distance from the Sun than when at its minimum distance (the Earth's orbit around the Sun is slightly elliptical).

• The angle of the Sun's rays due to the Earth's tilt - causes the equator and tropical regions to receive more sunlight than at higher latitudes.

• The reflectivity/absorption of the Earth's atmosphere – properties of atmospheric gases (including greenhouse gases), clouds, aerosols, ice, snow, sand, land, water, etc. reflect and/or absorb incoming radiation. Climate scientists use a quantity called *albedo* to describe the degree to which a surface reflects light; it is typically set at around 30% of incoming radiation.

This section comprises five subsections: The Earth's Energy Budget, the balance of incoming and outgoing energy (4.1.1); Solar activity; the variability of the Sun's brightness and energy transmitted to the Earth (4.1.2); Orbital parameters; the Earth's trajectory around the Sun and its planetary perturbations, tilt, wobble, and nutation, and (4.1.3); Planetary constellation of the largest planets, Jupiter and Saturn and its effect on the relative position of the Earth and Sun (4.1.4). Section 4.1.5 is a review of recent research.

To maintain a steady average global temperature, the Earth must emit the same amount of energy back into space as it receives from the Sun, i.e., the energy gained and lost are in balance; this is termed the Energy Budget.

4.1.1 Earth's Energy Budget: The energy entering, reflected, absorbed, and emitted by the Earth system are the components of the radiation budget. Based on the physics principle of conservation of energy, this radiation budget represents an accounting of the balance between incoming and outgoing radiation, which is partly reflected solar radiation and partly radiation emitted from the Earth. The unit of energy employed in measuring this incoming and outgoing radiation is watts per square meter (W/m^2). The ability of gases (atmospheric and greenhouse) to absorb radiation is a function of the chemistry between a gas and the intensity (wavelength) of the radiation with which it interacts. This is mostly important when, later in this chapter (section 4.6) we discuss greenhouse gas properties.

Depending on the source: IPCC, NASA, et al, the values assigned for the energy budget and specific energy flows varies, sometimes considerably. This is captured by the following NASA statements:

*"Determining exact values for energy flows in the Earth system is an area of ongoing climate research. Different estimates exist, and **all estimates have some uncertainty.**"*
(Earthobservatory.nasa.gov/features/energybalance), and in 2014 stated,

*"**Our understanding of those energy flows will continue to evolve** as scientists obtain new and longer records using new and better instruments."* (science-edu.larc.nasa.gov/energy_budget).

These are very important statements from NASA as will become clear in this chapter; errors or uncertainties in measurements often exceed the contributions of CO_2 on the energy budget. Figure 4.1-1 is an illustration of the components that affect the Earth's energy budget.

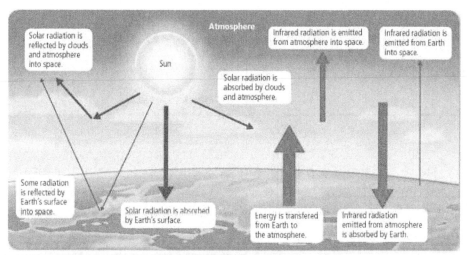

Figure 4.1-1 Earth's Radiation Budget System (Source: USGS)

Figure 4.1-2, which includes budget values, is a depiction of the energy budget elements from the 2007 IPCC Fourth Assessment Report (AR4), and is explained below. It uses a value of 342 W/m² as the average energy entering the atmosphere, which equates to 1,368 W/m² at the top of the atmosphere (4 x 342).

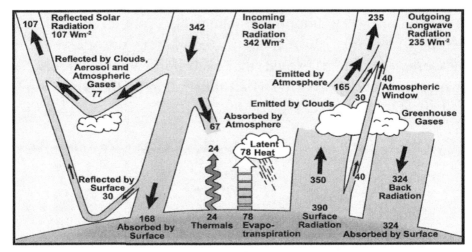

Figure 4.1-2 Earth's Energy Budget (Source: IPCC; AR4)

There are three separate energy balances: (1) at the top of the atmosphere (TOA); (2) within the atmosphere, and; (3) at the surface. Readers not interested in the calculations can skip the following three paragraphs (1, 2, and 3).

1. Energy entering the TOA = energy leaving the TOA
Across the top of the diagram shows the incoming radiation of 342 W/m² balanced by outgoing radiation: 107 W/m² reflected by clouds, aerosols, and atmospheric gases, plus 235 W/m² transported to outer space (165 from atmospheric gases + 30 from clouds + 40 that passes directly from the Earth's surface unhindered by the atmosphere). The cumulative amount of infrared radiation lost to space from the atmosphere: 107 + 235 = 342 W/m².

2. Energy gained by the atmosphere = energy lost to the atmosphere
Within the atmosphere itself (middle of the diagram), energy gained comprises 67 W/m² absorbed directly from incoming energy entering the atmosphere, plus 24 W/m² thermal heat from warm air rising through contact with the Earth's surface, plus 78 W/m² latent heat (evapotranspiration - evaporated heat from soil, plant life, vegetation, etc.), plus 350 W/m² radiated from the Earth

into the atmosphere where it is (temporarily) trapped by clouds and greenhouse gases for a total of 67+24+78+350 = 519. This is balanced by the atmosphere emitting 165 W/m into space, plus 30 W/m from clouds, plus 324 W/m of Back Radiation, i.e., heat returned to the Earth's surface from greenhouse gases (primarily water vapor, clouds and carbon dioxide), for a total of: 165 + 30 + 324 = 519.

3. Energy absorbed by the Earth's surface = energy leaving the surface (land and oceans)

The bottom of the diagram shows the amount of incoming energy absorbed by the surface is 168 W/m^2 which added to the back radiation from greenhouse gases of 324 W/m^2 = 492 W/m^2 total absorbed. This is balanced by the energy leaving the surface via thermals (24 W/m^2), evapotranspiration (78 W/m^2), and the surface emitting 390 W/m^2 of infrared (heat) radiation: 24 + 78 + 390 = 492 W/m^2.

While Figure 4.1-2 is satisfactory to illustrate the approximate energy budget flows, *specific* values vary considerably across research papers (recall AR5 used 1,361 W/m^2). And although improved satellite data gathering techniques and calibrations since 2007 claim more precision in measuring TSI and related energy flows, significant variances continue.

Table 4.1-1 shows IPCC AR4 values from Figure 4.1-2 and two others; a 2009 version from Dr. K. Trenberth, and a 2014 version from NASA. Dr. Trenberth is part of the Climate Analysis Section at the U.S. National Center for Atmospheric Research. He was a lead author of the 2001 and 2007 IPCC Scientific Assessment of Climate Change and contributed to the diagram shown as Figure 4.1-2. These data are from AGW advocacy sources.

Table 4.1-1 Selection of Research Article Results on Energy Budget

Source	Date	Effective TSI at the TOA W/m²	Incoming Radiation W/m²	Back Radiation W/m²	Net Surface Imbalance W/m²
IPCC AR4	2007	1,368	342	324	Not shown
Trenberth et al (1)	2009	1,365	341.3	333	0.9
NASA (2)	2014	1,362	340.4	340.3	0.6

(1) journals.ametsoc.org/doi/abs/10.1175/2008BAMS2634.1

(2) science-edu.larc.nasa.gov/energy_budget/pdf/

The disparity in Back Radiation (BR) is important to the climate discussion since BR represents heat that is absorbed by greenhouse gases and then returned to Earth; it is the *greenhouse effect*. The difference in BR values between the IPCC AR4 (2007) and NASA (2014) is 16.3 W/m². IPCC AR5, Chapter 8, Table 8.2, says that radiative forcing (i.e., the impact on climate) due to CO_2 is 1.82 W/m² ± 0.19, say 2 W/m², which is considerably less than the BR variation shown in the table. And, as will be shown in section 4.6, only about 5% of CO_2 is human-caused; i.e., the contribution of human-caused CO_2 is about 0.1 W/m² (0.05 x 2). Comparing this to the net energy budget contribution of clouds, around 28 W/m² puts the impact of CO_2 into perspective.

While IPCC AR5 (the most recent IPCC report) uses a TSI value of 1,361 W/m² in its models, Soon, Connelly and Connolly (2015, *Re-evaluating the role of solar variability on Northern Hemisphere temperature trends since the 19th century*) show that between 1978 and 2013 the variance among seven satellites recording TSI at the top of the atmosphere is around 14 W/m², the reason for which they say is unclear. This equates to a variance of 3.5 W/m² (14/4) entering the atmosphere, compared to the 0.1 W/m² attributed to human-caused CO_2; i.e., the **human-caused**

CO_2 contribution is about 3% of the *variance* in satellite measurements.

Another important data set shown in Table 4.1-1 is the "Net Surface Imbalance" which represents the excess heat absorbed by the Earth that must be emitted until energy balance is reached; a positive imbalance indicates a warming planet. Most AGW advocates ascribe this imbalance to human-caused CO_2. But note that it reduced by 30% from the 0.9 W/m^2 in 2009 (Trenberth) to 0.6 W/m^2 in 2014 (NASA) in spite of the continuing rise of CO_2 in the atmosphere.

A significant contributor to the radiation budget is the effects of *albedo*, i.e., the reflection of incoming sunlight from the atmosphere, clouds, aerosols, and the Earth's surface. It plays a vital role in the energy budget but is difficult to calculate with precision. It is typically estimated to reflect about 30% of incoming solar radiation, with about 22% reflected by clouds, aerosols, and atmospheric gases and the remainder from the surface (ice, snow, lightly colored sand, water, and other light surfaces such as concrete etc.).

Albedo, which theoretically can vary from zero (pitch black - no reflectivity) to 100% where all incident light is reflected (no absorption), has a substantial effect on the energy balance, and is highly variable. For instance, according to the website cited below, fresh snow has an albedo somewhere around 80 to 90%, forests have albedo near 15%, while the albedo of desert sands is roughly 40%. The website provides a model whereby users can alter albedo rates to find corresponding temperature variations. Using the model, and setting the albedo at 33% resulted in a global temperature decrease (more solar radiation reflected) of approximately 2°C, while decreasing albedo to 27% had a warming effect of about 3°C. These numbers, a ±10% sensitivity variation from a nominal 30%, result in a 5°C swing. Nobody suggests that a ± 10% variation in CO_2 levels has such an effect on temperature; IPCC models examine temperature variations based on doubling anthropogenic CO_2. **The Earth's energy balance, and therefore temperature, is far more sensitive to albedo** which is extremely difficult to measure since it is constantly varying, **than it is to CO_2.** (windows2universe.org/Earth/climate/sun_radiation_at_earth).

4.1.2 Solar Activity: Solar Activity is generally measured by the number of sunspots appearing on the surface of the Sun. According to NASA, sunspots are *"the most visible advertisement of the solar magnetic field"* and, as we will see, the Sun's magnetic fields and solar winds are a major influence on the Earth's climate.

The Sun is not a solid object as the Earth and Moon; it is a gaseous material and because of this it has a *differential rotation*; i.e., unlike the Earth which rotates at all latitudes every 24 hours, the Sun rotates every 25 days at the equator and takes progressively longer to rotate at higher latitudes, up to 35 days at the poles. This differential rotation warps internal magnetic fields which break through the surface creating sunspots, solar flares (brief bright eruptions of hot gas), and huge prominences (massive plumes of glowing gas that jut out of the Sun's surface that can extend 30,000 miles or more from the Sun's surface). Figure 4.1-3 shows NASA photographs from 2004 and 2017 of the Sun with sunspots (left) and clear of sunspots (center), and from a 1973 Skylab mission a spectacular solar prominence jutting out of the Sun's surface (right) and spanning more than 365,000 miles across the surface.

As can be seen from Figure 4.1-3, sunspots appear as dark areas on the surface and that is because they are in fact cooler regions than the rest of the surface.

Figure 4.1-3 Sunspot Cluster (2004); Free of Sunspots (2017); Right-Prominence (1973) (Source: NASA)

The relationship between sunspot cycles and TSI is more of a proxy as sunspots themselves are cooler regions and therefore would have reduced irradiance; they are dark spots with average temperatures of around 4,000°C. They are however, according to Gray et al (2010, *Solar Influence on Climate; Geophys.*), associated with the formation of other solar features called *faculae* which are brighter than the average surface, and the increase in solar irradiance from these faculae tends to outweigh the decrease in solar irradiance from cooler sunspots with a net increase in TSI. Faculae are less readily visible than sunspots because they are smaller, but they have a high surface temperature of around 6,000°C. Soon et al (2015) suggest that the increase in TSI during periods of high sunspot activity is mostly a consequence of the increase in facular emission, and not an increase in sunspots. But since sunspots are more easily counted they are an ideal proxy.

It is well recorded that it takes about 11 years for the Sun to move through a solar cycle, the time for it to rotate around the center of gravity of the solar system. Over these periods the number of sunspots varies from a (solar) minimum to a (solar) maximum and back to a minimum, which is then repeated. During these cycles, the magnetic material inside the Sun is constantly stretching, twisting, and crossing as it bubbles up to the surface.

The number of sunspots that occur during cycles varies considerably from none to a few to hundreds as shown in Figure 4.1-4. There have been 24 complete cycles recorded, the last one beginning in 2008 at solar minimum (and the least number of sunspots since the early 19th century), reaching its maximum in 2014 (with the lowest sunspot count since the beginning of the 20th century - 1902). Solar activity indicates a cooling trend.

According to the *Solar-Terrestrial Centre of Excellence*, December 2016, solar cycle 25 has started which it says doesn't mean that cycle 24 has ended since sunspot cycles usually overlap, up to 4 years.

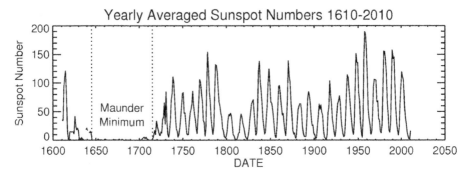

Figure 4.1-4 Sunspot Cycles/Numbers (Source: NASA/MSFC)

There have been times when the Sun appears to have ceased to develop sunspots for a lengthy period of time. Figure 4.1-4 shows that at the end of the 17th century to early 18th century (about 1645 to 1715-ish) there was a freezing period called the *Maunder Minimum* during which time there were virtually no sunspots. The River Thames in England froze over, and the first settlers to Jamestown, Virginia had the misfortune of arriving in that region during some of the coldest and driest weather of the Little Ice Age. Crop failures contributed to the horrendous mortality rates endured by the colonists, and to the temporary abandonment of their settlement.

A 24-year study by NASA, beginning in 1978 and extending over a notable global warming period, found that there was a significant positive trend (0.05% per decade) in TSI between the solar minima of solar cycles 21 to 23 (which also included two solar maxima cycles, 22 and 23).

According to Richard Willson, researcher and lead author of the study(nasa.gove/centers/goddard/news/topstory/2003),

"If a trend, comparable to the one found in this study, persisted throughout the 20th century, it would have provided a significant component of the global warming the Intergovernmental Panel on Climate Change reports to have occurred over the past 100 years."

In other words, if such a trend occurred during the 20th century, and there is no reason to believe otherwise, then much of the temperature increase over that period would have been caused by solar activity (leaving less warming to be attributed to CO_2).

There is demonstrably strong correlation between TSI and averaged sunspot cycles as shown in Figure 4.1-5, with some interesting results.

Figure 4.1-5 shows the estimates of TSI from numerous satellites launched since the beginning of the satellite era against three full sunspot cycles (22, 23, and 24). It was developed by William Ball, Joanna Haigh & National Center for Atmospheric Research Staff (climatedataguide.ucar.edu/climate-data/total-solar-irradiance-tsi-datasets-overview).

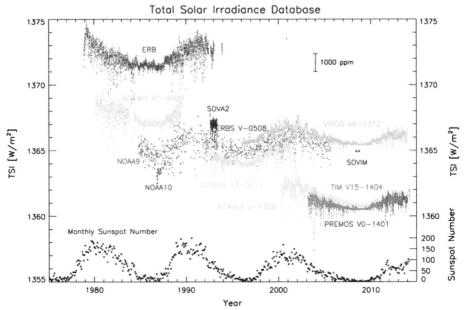

Figure 4.1-5 TSI versus Solar Cycles from 1978 to 2016
(Source: Public Domain: NCAR/UCAR)

It is clear from the figure that TSI varies in synch with sunspot cycles. Of particular interest however, is the variation between satellite-recorded TSI values during the same phase of a solar cycle. For instance at the solar maximum around 1980, during cycle 22, ERB and ACRIM1 satellites recorded TSI maximum values of around 1,375 and 1,369, respectively, a variance of 6 W/m². At solar minimum, around 1986, ERB recorded a TSI of about 1,371 while ACRIM1 recorded TSI at around 1,367, a 4 W/m² variance.

As discussed earlier, the effects of TSI on the Earth's average energy budget is one-quarter of these values, meaning that average variances of these data sets are about 1.5 W/m² during solar minimum and 1.0 W/m² during solar maximum. These numbers are significant when the net surface imbalance shown in Table 4.1-1, and a measure of global warming, is around 0.6 to 0.9 W/m², less than the variance between satellite measurements.

There is also a large body of research explaining potential *amplifiers* from the Sun's activity. For instance, NASA (science.nasa.gov/science-news/science-at-nasa/2013), suggests that the relatively small 0.1% variation in TSA during a solar cycle (this is the number used in IPCC models but we have seen it appears to be much larger) could have a measurable effect on climate through amplification that occur in synch with the 11-year solar cycles. These include (1) large variations in ultraviolet light emitted from the Sun that heats the stratosphere which, in turn, affects our (troposphere) climate, and (2) the effects of solar wind on galactic cosmic rays (that influence cloud formation). Vahreholt and Luning (2015) et al, also make strong cases for amplification effects of slight solar irradiation variance, also citing the Sun's influence on cosmic rays and the effects of ultraviolet light.

Ultraviolet (UV) Amplifying Effect: It should be non-controversial that changes in stratospheric UV affects the Earth's climate, or there would not have been a need for the Montreal Protocol which, in 1987 imposed radical changes on the use of hydrofluorocarbons and other chemicals that were shown to deplete the ozone layer which, in turn, allowed more harmful UV to penetrate the atmosphere.

There are three basic types of ultraviolet radiation, UC-A, UV-B, and UV-C. UV-A passes directly through the stratosphere and into the Earth's atmosphere, UV-B, the one primarily targeted by the Montreal Protocol, is partially absorbed by the stratosphere and, UV-C is totally absorbed in the stratosphere. The absorption of UV-B (partial) and UV-C cause the stratosphere to heat up as shown in Figure 4.1-6. Notice that unlike the troposphere in which temperatures decrease from around 15°C at the Earth's surface to about -60°C at the top of the troposphere, the temperature of the stratosphere increases with increasing altitude.

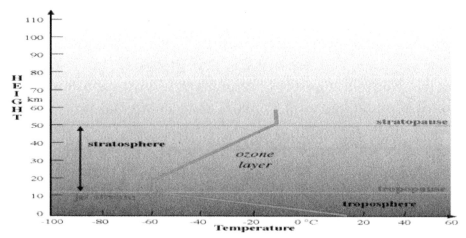

Figure 4.1-6 Temperature Profiles of the Troposphere and Stratosphere
(Source: Public Domain)

Gray et al, (2010) suggest that although the Sun emits very little UV radiation compared to visible light and infrared, UV *variability* is much greater during solar cycles when it can fluctuate from 5 to 10 times its nominal value. Vahreholt and Luning (2015) say that NASA's SORCE satellite measured UV radiation from 2004 through 2007 during the transition from solar maximum to solar minimum (cycle 24) and showed a change in UV to be five times greater than had previously been considered possible and that the temperature of the ozone layer during the solar maximum was almost 2°C higher than it was during the minimum.

Dr. Joanna Haigh, of London's Imperial College, presented a paper in the 1998 meeting of the American Association for the Advancement of Science (AAAS) on "*The Effects of Change in Solar Ultra-violet Emission on Climate,*" According to Dr. Haigh, a 0.1 % variation in TSI (the amount the IPCC says occurs during a typical solar cycle) could cause a 2% change in ozone concentration in the Earth's atmosphere that may lead to "*small but significant changes in the Earth's global weather patterns.*"

NASA's Shindell et al (1999, *Solar Cycle Variability, Ozone, and Climate, Science*) says his team confirmed that ozone is one of the key factors that amplify the effects of changes in the Sun's irradiance, and that there is evidence of atmospheric changes altering wind cycles, particularly at mid-to-upper latitudes with most effects on *Westerlies* and near the poles (see section 4.3 for more on wind effects on climate).

Chen Quanliang and Wei Lingxiao (2011; *Relationship of Temperature Variation between in the Stratosphere and the Troposphere*) note that the temperature variation in the stratosphere is very important in the global climate system and that it is considered to be the key to study atmospheric temperature vertical profiles (how the temperature varies throughout the total atmosphere). They suggest that the stratosphere can have an important effect on the weather systems in the troposphere through the Arctic Oscillation which, according to NOAA, is a climate pattern that produces colder air across Polar Regions (oceanic oscillations are discussed in section 4.4)

Finally, Edwin P. Gerber et al, (2012, *Assessing and Understanding the Impact of Stratospheric Dynamics and Variability on the Earth System*) suggests that there is conclusive evidence that the stratosphere plays a significant role in the natural variability and forced response of the Earth system. (journals.ametsoc.org/doi/pdf).

Cosmic Ray Amplifying Effect: Another, and probably the most important potential amplification factor that occurs during solar cycles, is the effect of solar wind and its magnetic field modulating Earth-bound cosmic rays.

• Solar wind is a stream of charged particles (plasma) released from the Sun's upper atmosphere, called the corona. Embedded within solar-wind plasma is the interplanetary magnetic field. The solar wind streams off the Sun in all directions at speeds of about 1 million mph.

• Cosmic rays (CR) are showers of high-energy radiation generated from exploding stars (supernova), mainly originating outside the solar system, and mostly from distant galaxies, that constantly bombard the solar system, and potentially the Earth. They travel almost at the speed of light.

Figure 4.1-7 shows images of solar wind (left) and supernova explosion (right). The strength of the Sun's magnetic field and solar wind determine the amount of CR that strikes the atmosphere. When the magnetic field is strong, it shields the Earth from the radiation. When it is weak, cosmic radiation penetrates deep into the atmosphere. The theory is that these cosmic particles seed clouds in the lower atmosphere and, as we will see in section 4.5, clouds significantly affect temperature. Increased solar activity reduces the amount of CR entering the atmosphere which leads to less cloud cover – warmer days; decreased activity results in more cloud cover – cooler days.

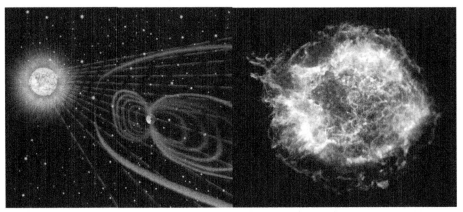

Figure 4.1-7 Left: Solar Winds vary in density, temperature and speed over time and over solar latitude and longitude. (Source: Public domain): Right: This photograph shows "Cassiopeia A," the youngest supernova remnant in the Milky Way (Source: NASA/CXC/MIT/UMass Amherst/M.D.Stage et al)

According to Dr. Nir Shaviv, clouds have been observed from space since the beginning of the 1980s. By the mid-1990s, enough cloud data accumulated to provide empirical evidence of a solar-cloud link. Without satellite data, it wasn't possible to obtain statistically meaningful results because of the large systematic errors plaguing ground based observations. This is very important because the IPCC charter casting human-caused CO_2 as the villain, occurred in 1988, before this research was possible. It has of course been dismissed – the tablets were already etched in stone! (sciencebits.com/CosmicRaysClimate).

Using satellite data, Dr. Henrik Svensmark (1998, "*Influence of Cosmic Rays on Earth's Climate,*" *Physical Review Letters*), of the Danish National Space Center in Copenhagen, reported that cloud cover varies in sync with the variable cosmic ray flux reaching the Earth. Over the relevant time scale, the largest variations arise from the 11-year solar cycle. In 2003, Svensmark and Marsh showed that the correlation is primarily with low altitude cloud cover. This can be seen in Figure 4.1-8 which, except for the outlier year, around 1999, shows remarkable synchronicity.

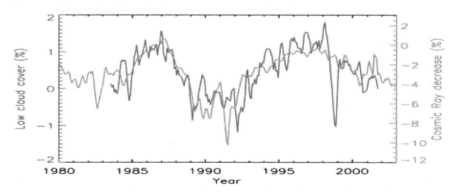

Figure 4.1-8 Cosmic Ray (red) correlation with Low-altitude Cloud Cover (blue)
(Source: Marsh & Svensmark, 2003)

A causal relationship however, has been challenged by some researchers reasoning that there is a possibility that solar activity modulates CRs and climate *independently*, and that the relationship may not be causal. For instance, A.D. Erlykin et al

(2009, *On the Correlation between Cosmic Ray Intensity and Cloud Cover; Journal of Atmospheric and Solar-Terrestrial Physics*), while apparently acknowledging a relationship between CR and low-cloud cover, cautions against attributing causal correlation to the effects. Instead they, and others, suggest the relationship is more likely to be *commensally correlated,* whereby the two data sets (cloud formation and CR variation) are independently correlated to another source; in this case to TSI variability during a solar cycle. In any case, the research shows that clouds are affected during a solar cycle, and clouds affect climate.

Over geological time, Jan Veizer, a geologist at the University of Ottawa, and Nir Shaviv, astrophysicist at the Hebrew University of Jerusalem, in a cross-disciplinary study published by the Geological Society of America (2003, *Celestial Driver of Phanerozoic Climate),* concluded that about 75% of the Earth's temperature variability in the past 500 million years was due to changes in the varying bombardment by cosmic rays as we pass in and out of the spiral arms of the Milky Way galaxy (the Milky Way has four spiral arms of varying matter density, resulting from its spin around a central axis), and that **less than half of the global warming seen since the beginning of the twentieth century is due to greenhouse gases**. They write,

*"Our approach, based on entirely independent studies from astrophysics and geosciences, yields a surprisingly consistent picture of climate evolution on geological time scales. **The global climate possesses a stabilizing negative feedback. A likely candidate for such feedback is cloud cover.**"*

In Figure 4.1-9, Veizer and Shaviv show reconstructions of cosmic ray flux (top panel) versus temperature (bottom panel) over the Phanerzoic Era.

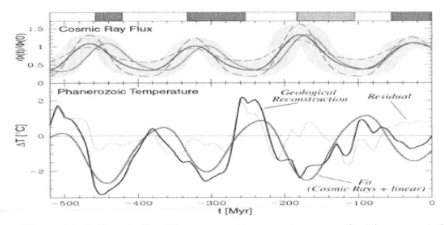

Figure 4.1-9: Cosmic Ray Flux versus Temperature over the Phanerozoic Era

The top panel shows the results of multiple simulations of reconstructed Cosmic Ray Flux (CRF) variations over the past 500+ million years. The bottom panel depicts the reconstructed tropical ocean temperature variations over this period (Geological Reconstruction) and the curve labeled Fit (Cosmic Rays + linear), the predicted temperature based on the CRF variations. The predicted temperature variation and empirical evidence from proxy data (bottom panel) show that there is a inverse relationship between CRF and temperature when measured over millions of years; i.e., **increased cosmic rays (more clouds) reduces temperature, decreased cosmic rays (less clouds) increases temperature.** They conclude that cosmic rays undoubtedly affect climate, and on geological time scales are the most dominant climate driver, saying,

*"The close fit implies that most of the temperature variations can be explained using the cosmic ray flux, and not a lot is left to be explained by other climate factors, including CO_2. This implies that **cosmic rays are the dominant (tropical) climate driver over the many million year time scale.**"*

Another key result of Dr. Shaviv's research (sciencebits.com/CosmicRaysClimate) is that variation of the CRF, as predicted from the galactic model and observed from proxy data, is also in sync with the occurrence of ice-age epochs; the observed period of the occurrence of ice-age epochs on Earth is 145 ± 7 million years, compared with 143 ± 10 for the CRF variations. Note, ice age epochs are different from glacial events which, as discussed in section 4.1.3, have a periodicity of about 100,000 years.

Veizer and Shaviv found little correlation between Earth's climate and carbon dioxide over the 500+ million years. They note that CO_2 levels have been 10 to 18 times higher during this period versus today's levels, and suggest that "*CO_2 is not likely to be the principal climate driver.*"

In a later article (2005, *Geosciences Canada* 32, no 1), Veizer found that "*empirical observations on all time scales point to celestial phenomena as the principal driver of climate.*"

"*Today,*" says Dr. Shaviv (2015, in *The Neglected Sun*), "*there is ample evidence linking solar activity to the terrestrial climate on all these timescales.*" And, "*Solar variations give rise to changes as large as 1 degree C between low and high solar activity.*" Adding, "*the leading contender to explain the solar-climate link is through cosmic ray flux and its affect on cloud cover.*"

Note, neither Veizer nor Shaviv exclude CO_2 as a *contributor* to climate change but that is not the main force. In fact they note that both CO_2 and solar/cosmic ray flux hypotheses need an *amplifier* to have a meaningful effect on climate. For CO_2 it is water vapor (discussed in section 4.6), and for solar/CRF it is clouds.

4.1.3 Orbital Parameters: The intensity and distribution of solar energy striking the Earth's surface varies with the Earth's changing orbit around the Sun, its axial tilt, and its wobble as it spins on its axis. Collectively, the total effect of these variations creates alterations in the seasonality and severity of solar radiation reaching the Earth's surface. These are called the Milankovitch Cycles, for Milutin Milankovitch (1879-1958), the Serbian astronomer and mathematician who is generally credited with calculating their magnitude. Each of these is discussed below.

Eccentricity: The first of the three Milankovitch Cycles is the Earth's *eccentricity*. Milankovitch asserted that the episodic nature of glacial and interglacial periods within the present Ice Age (the last couple of million years) have been caused primarily by cyclical changes in the Earth's circumnavigation of the Sun. The Earth's orbit approximates an ellipse. Eccentricity measures the departure of this ellipse from circularity.

This constantly fluctuating orbital shape, from almost circular to its most elliptical, has a cycle of about 100,000 years. The oscillation from less elliptic to more elliptic is of prime importance to glaciations in that it alters the distance from the Earth to the Sun, changing the distance the Sun's radiation must travel to reach Earth, subsequently reducing or increasing the amount of radiation (TSI) received at the top of the atmosphere (TOA).

The distance of the Earth from the Sun varies from about 91 to 94.5 million miles when nearly a circular orbit to about 80 to 116 million miles when at its extreme elliptical orbit. Today, with the orbit almost a circle, the difference in the amount of sunlight striking the Earth between extreme points of the orbit, from the nearest (perihelion – 91 million miles) to the farthest (aphelion – 94.5 million miles) is about 3.5%, which results in a variation in TSI of about 48 W/m² at the TOA between aphelion and perihelion during an annual cycle. This computes to about 12 W/m² in the atmosphere. Recall from the earlier discussion on the Earth's Energy Budget, that the contribution of CO_2 to the energy budget is about 2 W/m². **Orbital differences, which have six** times more effect on the Earth's energy budget than CO_2, are considered negligible by the IPCC et al.

When Earth's orbit is most elliptical (80 to 116 million miles), the amount of solar energy received at the perihelion would be in the range of 20 to 30% more than at aphelion, a variation from about 272 W/m^2 to 408 W/m^2. Additionally, a more eccentric orbit will change the length of the seasons by changing the time between spring and autumn equinoxes; the speed of the Earth's rotation is fastest when it is closest to the Sun and slowest when it is farthest away from the Sun. Figure 4.1-10 depicts Earth's orbital extremes.

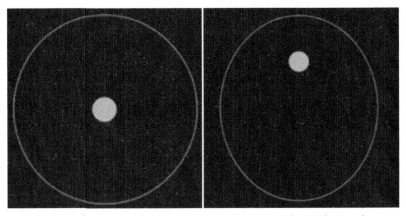

Figure 4.1-10: Left, Circular orbit, no eccentricity; Right, Orbit with extreme eccentricity (Image by Robert Simmon, NASA GSFC)

The extreme excursions of the orbit have been attributed as the cause of the 100,000-year glacial/interglacial periods, shown in Figures 3-1 and 3-2 in Chapter 3. According to *Science Daily* (2013),

"Ice ages and warm periods have alternated fairly regularly in Earth's history: Earth's climate cools roughly every 100,000 years, with vast areas of North America, Europe and Asia being buried under thick ice sheets. Eventually, the pendulum swings back: it gets warmer and the ice masses melt."

Only changes in the eccentricity lead to changes in the absolute amount of solar radiation received by the Earth. Apart from variations due to solar activity, the Sun emits the same amount of radiation regardless of the Earth's orbital position, but the amount of radiation reaching the Earth is determined by the distance between them; the intensity of radiation varies with the inverse square of the distance from the source (for example, if the Earth was twice its current distance from the Sun, it would receive ¼ the amount of radiation). Obliquity (tilt) and precession (wobble) affect the relative *distribution* of the incoming radiation. It is the distribution of energy, warmer at the tropics, cooling towards the Poles, that creates atmospheric and oceanic circulations (discussed in section 4.3) and climate variation.

Obliquity: The second of the three Milankovitch Cycles is the inclination or *tilt* of the Earth's axis in relation to its plane of orbit around the Sun. The angle of the axial tilt with respect to the orbital plane varies between 22.1° and 24.5° over a cycle of approximately 41,000 years. The current tilt is about 23.4°, roughly halfway between its extreme values. Figure 4.1-11 illustrates the tilt range; the vertical line on the right indicates the orbital plane.

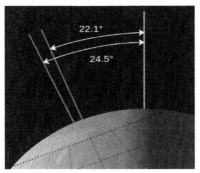

Figure 4.1-11: 22.1–24.5° Range of Earth's Obliquity
(Image by Robert Simmon, NASA GSFC)

The tilt determines how directly we receive radiation from the Sun; the intensity of the sunlight striking the Earth varies from maximum to minimum with latitudes from the tropics to the poles, respectively. On an average yearly basis, areas north of the Arctic Circle receive only about 40 percent as much solar radiation as equatorial regions.

Because of the periodic variations of this angle, the severity of the Earth's seasons changes. With less axial tilt solar radiation is more evenly distributed between winter and summer. **If there was no tilt there would be no seasons**. The effect of this can be seen during equinoxes when the tilt has no effect; i.e., when the tilt of the Earth's axis is inclined neither away from nor toward the Sun. Figure 4.1-12 depicts the distribution of the Sun's energy during an equinox, and as it would be if there was no tilt.

Equinoxes are the times of the year when the Sun's most direct rays hit the equator; currently they occur in March (spring) and September (autumn). Both hemispheres receive the same amount of energy and the length of the day and night is about 12 hours each, everywhere on Earth.

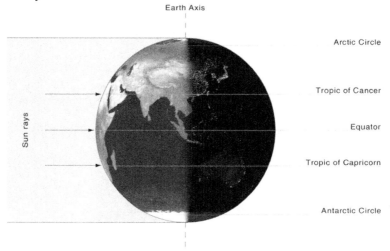

Figure 4.1-12: Distribution of sunlight during equinoxes
(Source: Public domain)

Seasons are a direct consequence of the Earth's orbit and tilt; with the orbit nearly circular, tilt has a more profound effect on the seasons. The direction of the Earth's axis stays nearly fixed throughout one orbit so that at different parts of the orbit, one hemisphere tilts or leans towards the Sun, while the other leans away.

Figure 4.1-13 illustrates the relationship between the orbit and tilt. At the June Solstice, the Earth tilts toward the Sun such that the N-H receives more direct sunlight (summer); at the same time the southern hemisphere (S-H) tilts away from the Sun (winter). Six months later at the December Solstice, the opposite situation occurs; the N-H leans away from the Sun (winter) while the S-H leans toward the Sun (summer).

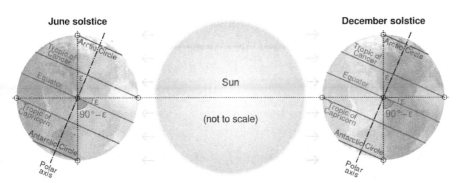

Figure 4.1-13: Axial Tilt, Seasons, and Solstices
(Source: cmglee, NASA, Public Domain, via Wikimedia Commons)

At the June Solstice, N-H summer, the Sun is at its most distant position of its orbit from the Earth (aphelion), but the tilt causes it to receive more energy from the Sun than when it is closer during winter. The S-H is closer to the Sun in its summer than it is in winter, i.e., the S-H summer occurs at perihelion receiving about 3.5% more radiation, so it would seem obvious that S-H summers are warmer than N-H summers. This is not the case however, because land warms (and cools) faster than water, and the S-H is about 80% water and 20% land, whereas the N-H is about 40%

water and 60% land. As a result of tilt and the land-to-water ratio (and an orbit that is nearly circular), N-H summers and winters are more extreme, while the S-H climate has less variation between seasons.

At the June and December solstices, the Sun's most direct rays strike the Tropic of Cancer and the Tropic of Capricorn, respectively. This is depicted in Figure 4.1-13 which also shows why temperatures cool with increasing latitude, from the equatorial tropical regions to the Poles. It's a result of the Earth's (almost) spherical shape and tilt causing the Sun's energy to be spread over a larger area so that its intensity (W/m²) is reduced.

The left side of the figure shows the configuration of the Earth and Sun during a N-H summer when the Earth tilts toward the Sun, i.e., at the June solstice (around 21-22 June). Clearly, the N-H, from the top of the orbital plane (vertical line) to the equator receives more sunlight than the portion of the S-H from the equator to the bottom of the Earth's orbital plane. The most sunlight at this time is received at the Tropic of Cancer which is the closest point to the Sun.

Six months later (right side of the figure), during the December solstice (21-22 December), the opposite situation occurs, with the S-H, from the Equator down past the Tropic of Capricorn (which receives the most direct rays at this time) to the bottom of the Earth's orbital plane receives more direct sunlight than the N-H from the Equator up through the edge of the Arctic Circle.

The differences in temperature distribution cause winds which, together with oceanic circulation, moves heat from equatorial latitudes toward the poles (discussed in section 4.3).

Precession: The third aspect of the Milankovitch Cycles is precession. Precession is the Earth's slow wobble as it spins on its axis. Since the Earth is a spheroid, not a sphere (it is roughly 13 miles greater distance to the center of the Earth at the equator than at the poles), as it rotates on its axis centrifugal force causes the equator to bulge. The non-uniform gravitational force of the Sun and the Moon pull on this bulge and causes it to wobble as it spins around it axis. Precession of Earth's rotational axis takes approximately 26,000 years to make one complete revolution. Through each 26,000-year cycle, the direction in the sky to which the axis points go around a big circle. Currently our axis in a northern direction is pointing at Polaris, i.e., the North Star. Due to precession, in about 13,000 years the axial direction it will point at the star Vega, which will then be the North Star. In another 13,000 years it will again point toward Polaris. This process, shown in Figure 4.1-14, modulates the effects of the previous two cycles in varying the seasons.

Figure 4.1-14: Precessional Movement
(Image by Robert Simmon, NASA GSFC)

When Vega becomes the North Star, summers and winters would be on the opposite sides of the orbit. Instead of winter in the northern hemisphere occurring during the part of the orbit when we are closest to the Sun (perihelion), it would be at the farthest point (aphelion). Northern hemisphere winters would be colder, but summers would be warmer. In addition to the Milankovitch cycles there is another astronomical influence on the season, *nutation*, caused by the orbits of the Sun and Moon.

Nutation: Nutation is a rocking, swaying, or nodding motion in the axis of rotation during precession, the result of which is a small change to the tilt angle (obliquity), which causes (relatively) short term disturbance to the seasons and to tidal behavior. The principal cause of nutation are interactions between the orbits of the Sun and Moon, which continuously change location relative to each other and disturb the attitude of the Earth's axis. According to Plimer (2009, *Heaven and Earth*) et al, over a period of 18.6 years the Moon's trajectory moves farther north over the equator and then back again due to the Sun's gravity. This has such an effect on tides that, in 2006 for instance, tidal currents brought large volumes of warm water into the Arctic, the Arctic heated, the summer ice melting rate increased, and some glaciers started to thin. Plimer says that the mass media promoted the idea that the decreased ice was proof of human-induced global warming when in fact it was a natural response to the alignment of the Earth, Sun, and Moon within the 18.6-year cycle. Figure 4.1-15 depicts nutation modulating the Earth's obliquity. Without nutation (shown as oscillating about the precession orbit), precession would follow the smooth path indicated by the dotted line.

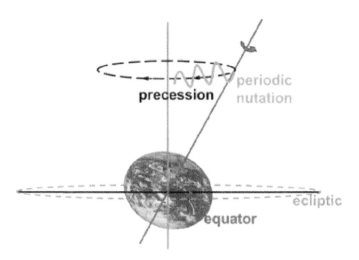

Figure 4.1-15 Periodic Nutation affecting Precession
(Source: public domain)

Precession and nutation are caused by the gravitational forces of the Moon and Sun acting upon the non-spherical Earth. The reason for including nutation in addition to the Milankovitch cycles is to demonstrate another astronomical force that potentially affects climatic conditions.

4.1.4 Planetary constellations: As mentioned above the gravitational effects of other planets, primarily Jupiter and Saturn, are responsible for the Earth's variable orbit around the Sun and its associated effect on the amount of solar radiation striking the Earth. According to Scafetta (2010, *Journal of Atmospheric and Solar-Terrestrial Physics*), there is a 60-year astronomical cycle that may be associated with oceanic oscillations, specifically the Pacific Decadal Oscillation (PDO) and North Atlantic Oscillation (NAO). As we will see in section 4.3, PDO cycles have a strong correlation with temperature although, because of the 60-year cycle period, there aren't sufficient samples to draw firm conclusions, and NAO plays a major role in weather and climate variations over large regions of the globe.

Jupiter takes about 12 years to orbit the Sun while it takes Saturn about 30 years. This means that every 20 years or so the two planets meet and align with the Sun. During such alignment, the planets pull most strongly at the Sun, while this force weakens as the two planets transit into different parts of the solar system. After three of these so-called conjunctions that is, about 60 years, Jupiter and Saturn occur approximately in the same constellation. Therefore, about every 60 years there will be a convergence where the two planets are closer to the Sun than during the previous two events; their gravity pull affects the Sun's position relative to Earth which affects the amount of radiation received on the Earth which in turn, affects the climate.

4.1.5 Current research: According to Kenneth Richard, (Blog, 12, January 2017), scientists are increasingly tuning out the claims that the Earth's temperatures are predominantly shaped by anthropogenic CO_2 emissions. Instead, solar scientists continue to advance their understanding of solar activity and its effect on the Earth system, and their results are progressively suggesting robust correlations between solar variability and climate change.

For example, in 2016 alone, there were at least 132 peer-reviewed scientific papers documenting a significant solar influence on climate. Among them there were 18 papers that directly connect centennial-scale periods of low solar activity with cooler climates (e.g., Little Ice Age), and periods of high solar activity with warm climates (e.g., Medieval Warm Period and the Modern Warm Period of the 20th/21st Century). Another 10 papers warn of an impending solar minimum and concomitant cooling period in the coming decades. The trend of scientists linking climate changes to solar forcing mechanisms continues to grow. See more at: http://notrickszone.com/2017/01/12/scientists-find-climates-cause-of-causes-highest-solar-activity-in-4000-years-just-ended-cooling-begins-in-2025.

According to Vahrenholt and Luning (2015), there are literally hundreds of studies covering periods from the last glacial, about 150,000 years ago, through the 20th and early 21st centuries that show strong correlations between temperatures and solar activity. Most of the studies cited and discussed are about direct contributions of solar activity with temperature, sea levels, etc, whereas other climate scientists, such as Spencer, Scafetta, et al suggest indirect solar influence via its effect on oceanic oscillations such as the PDO. And Professor Bond et al (*Science*, 2001) showed strong correlations between solar activity and the northern Atlantic climate over the past 10,000 years.

Many findings show correlations between different climate reconstructions (proxies) and different solar activity for all regions of the Earth, not just Europe as is often claimed by AGW advocates. For instance, Neff et al (2001, *Nature*, 411) showed a multi-millennial correlation between solar activity and temperatures of the Indian Ocean. And a more recent research paper (February 2017, *Science*), studying the effects of solar forcing on Indian summer monsoons, concluded that there is a good match between the proxy data and the reconstructed total solar irradiance data for the past 1,200 years, and that *"the Sun exerts a profound   control   on   the   southwest   monsoon."* (sciencedirect.com/science/article)

<u>Section 4.1 Summary:</u> **Compared to the effects of the Sun, CO_2 is a minor disturbance on the Earth's energy balance.** The laws of physics require that the amount of energy entering the Earth's system equal the amount of energy exiting the system. When an imbalance occurs due to one or more climate forcing events, the Earth's systems react in such a manner to restore the balance. If this were not the case, the temperature would spiral in one direction or another, and of course it doesn't.

Section 4.1.1, Earth's Energy Budget, showed that the calculated imbalance claimed to be responsible for the Earth's warming (0.6 to 0.9 W/m^2) is dwarfed by the variation in satellite measurements of the Sun's radiation (3.5 W/m^2 at the Earth's surface), and this variation is about 96% greater than that attributed to the human-caused CO_2 contribution (3.5 W/m^2 versus 0.1 W/m^2).

Section 4.1.2, Solar Activity, demonstrated the effect of solar activity on the Earth's radiation budget as measured by the 11-year solar cycle sunspot count. It appears that sunspots themselves are essentially proxy data for other solar effects. A lead researcher for a NASA study on the relationship between sunspot count and the amount of radiation reaching the Earth noted that if such a trend existed during the 20th century then much of the temperature rise over that period could be attributed to solar activity (thereby reducing the impact of CO_2). This section also discussed amplification effects of solar activity by (1) significantly increasing the level of ultraviolet in the stratosphere affecting weather patterns and climate in the troposphere, and (2) its magnetic field modulating Earthbound cosmic rays and the formation of low clouds which, in turn, affect climate.

Section 4.1.3, Orbital Parameters, describes the impact on climate resulting from (1) its varying orbital path around the Sun which effect the total amount of radiation received at the top of the atmosphere; (2) its tilt in relation to the plane of the Sun which effects the distribution of sunlight – more intense at the tropics attenuating toward the poles as the same amount of sunlight is dispersed over a larger area; (3) its "wobble" caused by the gravitational pull of the Sun and Moon which affect the seasons. The section also discussed another astronomical event, *nutation*, and its 18.6-year cycle of the Moon's trajectory in relation to the Sun that impacts tidal behavior which affects northern ocean temperatures (Arctic).

Section 4.1.4, Planetary Constellations, when the two largest planets Jupiter and Saturn occur in the same constellation, about every 60 years, they coincide with Pacific and Atlantic oscillations that have demonstrated climate impacts.

In spite of the evidence in this section showing clearly that the Sun has enormous effect of the Earth's climate, the IPCC effectively ignores it as a major climate-change force.

4.2 Tectonics:

Over *geologic time* the Earth's continents and oceans have shifted radically and most importantly for this discussion, they continue to move as much as 8 inches a year. Figure 4.2-1 shows various stages of continents/oceans evolution over the past 225 million years, from a single *super* continent, Pangaea, to today's configuration.

This shifting of continents and oceans is the result of plate tectonics. Although various tectonic theories had been proposed for more than a hundred years, it was not until the 1960s that geologists began to accept that continents did, and continue, to move, and it was not until satellites and GPS technology arrived that continental *drift* could be accurately measured. This section begins with a primer on plate tectonics theory, followed by discussion on some of the ways that movement of continents and oceans affect climate.

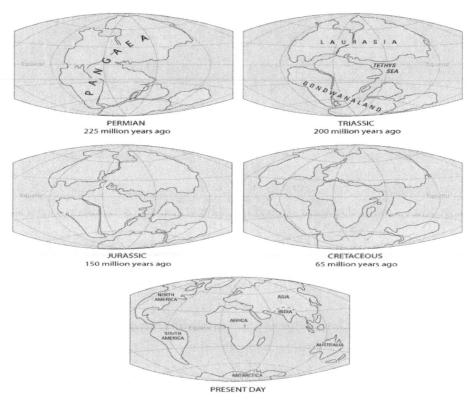

Figure 4.2-1 Stages of Continental Shifting from Pangaea to today
(Source: USGS)

The Earth consists of three layers: crust, mantle, and core (Figure 4-2-2). Each of these layers has two main parts: the crust is the surface layer of the Earth and is subdivided into two components, oceanic and continental (land). Oceanic crust is denser and thinner than continental crust. Oceanic crust is about 3 miles thick versus up to 40 miles thick for continental crust. The outer part of the mantle, on which the crust rests, is solid rock while the inner part is highly viscous molten rock, heated by the decay of heavy metals, such as uranium. The thickness of the mantle is about 1,800 miles. The core comprises a liquid outer segment and a solid inner section; its thickness is around 2,200 miles.

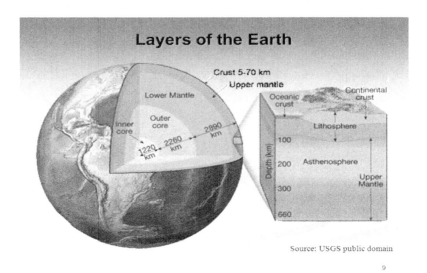

9

Figure 4.2-2 Layers of Earth (Source: USGS)

The crusts plus the outer solid mantle combined is called the *lithosphere*. The lithosphere *floats* on the hot, viscous fluid part of the mantle, called the *asthenosphere*. The lithosphere is a fractured shell broken into sections called plates. Plate tectonics is the theory that the Earth's lithosphere, floating on the viscous part of the mantle, the asthenosphere, is in constant motion.

Lithosphere plates include both continental and oceanic crusts; they differ in size and direction of motion. According to Chamberlin and Dickey (2009, *Exploring the World Oceans*), there are 14 major plates: the <u>African</u>, <u>Antarctic</u>, Arabian, <u>Australian</u>, Caribbean, Cocos, <u>Eurasian</u>, Indian, Juan de Fuca, Nazca, <u>North American</u>, Pacific, Philippine, and <u>South America</u>, and 38 minor ones. Six of the major plates (underlined) are named after the continents they contain. The biggest plate, the Pacific plate, is the only tectonic plate that is mainly under water (it has a tiny land crust at the edge of California), although it is shrinking as the Atlantic Ocean increases in size pushing the North American Plate slowly westward.

Figure 4.2-3 shows the major plates and their boundaries. Plates are continuously moving; where plates meet their boundaries, they will converge (crush into each other), diverge (separate from each other), or transform (grind past each other). The boundary geography is indicated by the bold black lines and their type by the direction of the arrows. It takes a bit of effort to observe the plate boundaries and the land and ocean areas contained within plates but you can see for example, the Eurasian plate includes Europe and Asia land as well as part of the North Atlantic Ocean. Its boundary with the North American plate is divergent, while the boundary with the Indian plate is convergent. The North American plate includes North America, Greenland, and the rest of the North Atlantic Ocean; it has a divergent boundary with the Eurasian plate, a convergent boundary with Pacific plate (north and north-west boundaries), and transform boundary along the Californian coastline.

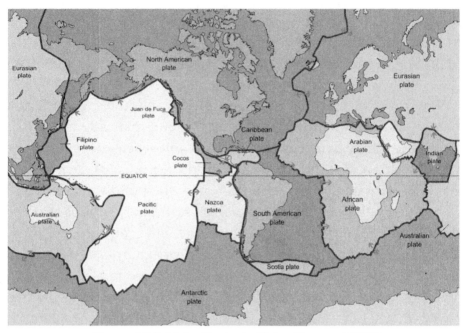

Figure 4.2-3: Major Tectonic Plates (Source: USGS)

Convergent Boundaries: Convergent boundaries can be broken down into three subcategories; continent-continent; ocean-continent, and; ocean-ocean. The effects of each of these collisions are different because oceanic crusts are denser and thinner than continental crusts, and older oceanic crusts are denser than newer oceanic crusts. Upon collision, a denser plate will dive below a less dense plate pushing the latter upwards; this is called *subducting*.

Continent-continent: When two continent plates converge, because they have similar density neither plate subducts, so they buckle and force the land between them to rise and create mountain ranges as depicted in Figure 4.2-4. For example, the Himalaya mountain range was formed by the collision of the Eurasian and Indian plates. Figure 4.2-5 shows the direction of convergent plates (dark black Arrows) and the location of the Tibetan Plateau relative to the Himalayas. The effect of this relationship on climate is discussed later in this section. The Himalayas continue to rise by about ½ inch per year as the plates continue to collide.

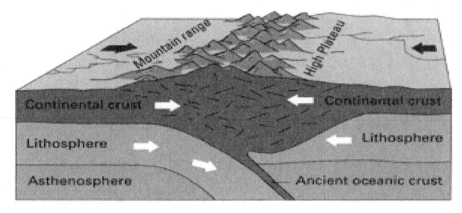

Figure 4.2-4 Continental – Continental Convergence (Source: USGS)

Figure 4.2-5 The Collision between the Indian and Eurasian plates pushed up the Himalayas and the Tibetan Plateau (Source: USGS)

Ocean - Continent: Oceanic crust is denser and thinner than continental crust, so the oceanic crust gets bent and subducted beneath the lighter and thicker continental crust, as depicted in Figure 4.2-6. Subduction zones have a lot of intense earthquakes and volcanoes as well as long, narrow trenches extending thousands of miles and up to 6 miles deep cutting into the ocean floor. The ranges of volcanoes, which occur in the continental crust, are known as a volcanic arc. Volcanoes are caused by molten rock, resulting from the friction and pressure caused by the subduction, rising through the lithosphere and blowing through the surface of continental crust.

Off the coast of South America along the Peru-Chile trench, the oceanic Nazca Plate is pushing into the continental part of the South American Plate (Figure 4.2-3) where it subducts. The South American Plate is lifted up, creating the towering Andes Mountains, the backbone of the continent. Strong, destructive earthquakes and the rapid uplift of mountain ranges are common in this region.

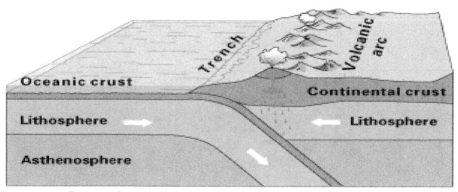

Figure 4.2-6 Oceanic-Continent Convergence (Source: USGS)

Ocean – Ocean: When two plates carrying oceanic crusts collide, the older one, which has become denser over time, subducts below the younger one. This is depicted in Figure 4.2-7 with older ocean crust shown on left. The effects are deep earthquakes, an oceanic trench, and a chain of volcanic island arcs within the ocean. Japan, the Aleutian Islands of Alaska, and Indonesia were formed by such collisions. **The 2004 Indonesian earthquake that** produced the devastating tsunami and the 2011 earthquake and tsunami that struck Japan were created by this type of collision.

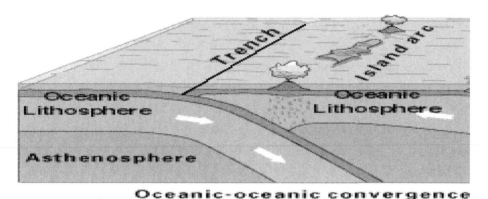

Figure 4.2-7 Ocean-Ocean Convergence (Source: USGS)

Divergent Boundaries: Oceanic ridges occur at divergent boundaries where new seafloor is produced (Figure 4.2-8).

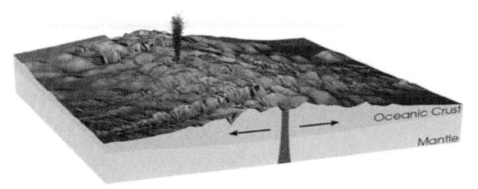

Figure 4.2-8 Divergent Boundary and the formation of Oceanic Ridges (Source: USGS)

The gap caused by divergence is filled from below by volcanic activity upwelling magma from the mantle, which cools upon reaching the ocean floor. This newly formed rock at the ocean floor pushes the existing crust away from the gap, initially creating, and then adding to, a ridge or mountain range. This is called *seafloor spreading*.

Volcanic eruptions occur mostly along ridge boundaries creating new ocean floor and at the same time pushing the two tectonic plates apart at rates of, according to Chamberlin and Dickey (2008), 0.5 to as much as 8 inches per year. According to Plimer (2009, *Heaven and Earth*), about 85% of the planet's volcanoes are submarine and account for 75% of the heat transferred to the surface from molten rock, and when molten rocks rise and cool, they release *"monstrous amounts of CO_2"* and other gases and, **while visible volcanoes represent only 15% of the Earth's total, they are the only volcanoes accounted for in climate models.**

When the divergent process begins, a valley will develop. Over time that valley can fill up with water creating linear lakes (long narrow line). If divergence continues, a sea can form like the Red Sea in northeastern Africa. Figure 4.2-9 is a Google Earth satellite image focusing on the Red Sea. It shows the Red Sea with a rift valley along the center, which used to be a linear lake that grew into a sea. If divergence continues and the Red Sea continues to grow it could eventually form an ocean similar to the Atlantic Ocean; the ultimate divergent boundary is the Atlantic Ocean, which began when Pangaea broke apart.

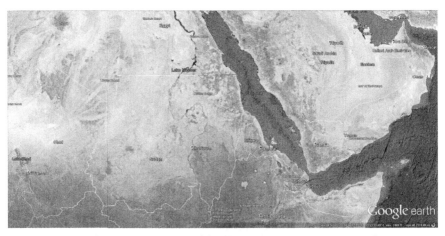

Figure 4.2-9 Red Sea formed by Tectonic Divergent Boundaries
(Source: Google Earth)

The Mid-Atlantic Ridge is the longest and the most extensive chain of mountains on Earth; it connects with other oceanic ridges to form a nearly continuous subsurface mountain range network more than 30,000 miles long, which plays a vital role in oceanic circulation, as discussed in section 4-3.

According to Chamberlin and Dickey, the rate at which the Mid-Atlantic Ridge expands averages about 1 inch per year, or 16 miles in a million years. This rate may seem slow, but over the past 100 to 200 million years it has caused the Atlantic Ocean to grow from a tiny inlet of water between the continents of Europe, Africa, and the Americas into the vast ocean that exists today.

Transform Boundaries: Transform boundaries occur when two tectonic plates slide or grind parallel to each other. This can be in the same direction where plates move at different rates, or in opposite directions. Although transform boundaries are not marked by spectacular surface features, their sliding motion causes earthquakes. Probably the most well-known transform boundary is the San Andreas Fault where the Pacific Plate (that Los Angeles and Hawaii are on) is grinding, in the same direction, past the North American Plate (that San Francisco and the rest of the United States are on) at a rate of about an inch a year. So, in a few million years, Los Angeles and San Francisco will be next to each other.

Climatic Effects of Plate Tectonics: On December 26, 2004, one of the most powerful earthquakes in the last 100 years occurred off the coast of Indonesia creating a massive tsunami killing more than 240,000 people. In October 2005, a powerful earthquake in Kashmir, India killed over 80,000 people. And on January 12, 2010, an earthquake devastated the Caribbean nation of Haiti, leaving more than 250,000 dead, 300,000 wounded and more than one million people homeless. On March 11, 2011, Japan was rocked by a magnitude 9.0 earthquake followed by a devastating tsunami that killed over 30,000 people and created a nuclear disaster. **These were natural disasters caused by plate tectonics that had nothing to do with anthropogenic CO_2.**

Figure 4.2-1 showed that North and South America were not always connected. According to a NASA web site (March 23, 2008), twenty million years ago, where Panama is today, there was a gap between the continents of North and South America, through which the waters of the Atlantic and Pacific Oceans flowed freely, balancing the salinity of the world's oceans which, as discussed in section 4.3, controls oceanic circulation. Beneath the surface, two plates of the Earth's crust (Cocos and Caribbean) slowly collided into one another. The pressure and heat caused by this collision led to the formation of underwater volcanoes, some of which grew tall enough to break the surface of the ocean and form islands; this was about 15 million years ago. More and more volcanic islands filled in the area over the next several million years. Meanwhile, the movement of the two tectonic plates also pushed up the sea floor, forcing some areas above sea level. By about 3 million years ago, a narrow strip of land had formed connecting North and South America. This was the formation of the Isthmus of Panama.

The NASA website says that scientists believe the formation of the Isthmus of Panama is one of the most important geologic events to happen on Earth in the last 60 million years, and that even though it is only a tiny sliver of land relative to the sizes of continents, **the Isthmus of Panama had an enormous impact on Earth's climate and its environment.**

By shutting down the flow of water between the two oceans, the land bridge re-routed currents in both the Atlantic and Pacific Oceans. Atlantic currents were forced northward, and eventually settled into a new current pattern, one of the outcomes of which was the Gulf Stream which made Europe and other northwestern regions habitable. The Atlantic, no longer mingling with the

Pacific also grew saltier impacting oceanic circulation patterns which transport heat and precipitation around the Earth. The movement of continents, the creation of mountains, new islands, etc., also affect atmospheric and oceanic circulation patterns.

Atmospheric: As discussed earlier, the Himalayas were formed when the Indian and Eurasian plates collided (Refer to Figure 4.2-5). One way that mountains affect weather is by creating rain shadows, i.e., by forcing warm, moist air up to altitudes where it cools and condenses causing heavy rains on the windward side of the mountain, leaving dry air to settle on the leeward side of the mountain. Figure 4.2-10 illustrates this process.

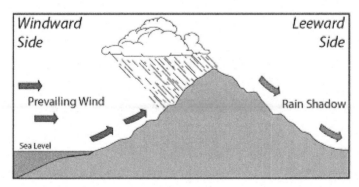

Figure 4.2-10 Illustration of Rain Shadow Effect (Public Domain)

In summer, winds blow northward off the Indian Ocean producing hot, humid air, but the Himalayas block these moist winds, creating a huge rain shadow. The moisture is squeezed out of the air over southern Asia producing heavy rain – the windward side of the Himalayas receives 1,500 to 3,000 mm (about 60 to 120 inches) of rain annually. Meanwhile because of the shadow effect, the Tibetan Plateau north of the Himalayas stays bone dry. The Himalayas continues to rise by about ½ inch per year.

Another example of the rain shadow effect of mountains is the Mojave Desert which lies in the rain shadow of the Sierra Nevada

Mountains in the USA. The hot, moist air from the Pacific Ocean rises up the Sierra Nevada's where it cools and rains. Death Valley, located in the Mojave Desert is one of the hottest and most arid places on Earth.

According to Randall (2012, *Atmosphere, Clouds, and Climate*), the major mountains ranges of the world cause prominent wave patterns in global circulation which influence the patterns of atmospheric heating and cooling.

Oceanic Circulation: Subsurface mountains or ridges are formed by oceanic-oceanic plate divergence. For example (refer to Figure 4.2-3) the Mid-Atlantic Ridge was formed by the divergence of the North American plate moving westward, and the Eurasian and African plates moving eastward. The ridge bisects the Atlantic Ocean (it also bisects Iceland above the surface) and plays a significant role in the flow of the ocean conveyer belt discussed in section 4.3. Mid-oceanic ridges of the world are connected to form a single global undersea mountain range more than 30,000 miles long weaving around the oceans. These subsurface mountains or ridges, which continue to evolve, directly affect the flow of the Ocean Conveyer Belt (discussed in section 4.3) which moves heat and moisture around the globe.

Section 4.2 Summary: Tectonics has, over geologic time, altered the entire structure of the Earth's landmass and ocean floor topology. While it may be convenient for advocates of the AGW CO_2 hypothesis to discount events measured in geologic time from climate change discussions, it is important to understand that plates *continue* to move, mountains *continue* to rise, undersea ridges and ocean seafloor spreading *continue* to occur; earthquakes, volcanoes, and tsunamis are the manifestation of these continuous movements. Plate tectonics continues to evolve the structure of the planet slowly changing atmospheric and oceanic circulation and their attendant effect on climate.

4.3 Wind and Ocean Circulation:

The source of wind is the Sun. As we have seen in section 4-1, the Sun's energy is unevenly distributed over the Earth with maximum radiation at tropical regions, where its rays strike most directly, attenuating towards the poles as the angle of incidence causes the Sun's energy to be distributed over a wider area and therefore less intense. This causes transfers of energy that result in wind and ocean circulations.

Wind and ocean circulations are coupled in a complex interactive system that affects regional and global climate. The combination of wind and ocean circulations redistributes heat and moisture across the Earth's surface and throughout the Earth system; they have a moderating effect on the global distribution of temperature, tending to warm the higher latitudes and cool the tropics. Around 60% of the solar energy that reaches the Earth is redistributed around the planet by atmospheric circulation and around 40% by ocean currents.

Wind moves heat, clouds, and moisture around the globe. Wind drags on the ocean surface creating surface currents, such as the Gulf Stream, which affects climate. Wind move warm surface water from the equator towards the poles where it cools and sinks to the deep ocean. In the equatorial Pacific Ocean winds, moving between the eastern edge of the Pacific and the west, create such events as El Niño/La Niña which have major climate impacts. Deep Ocean currents transport warm water and precipitation from the equator toward the poles and cold water from the poles back to the tropics.

Wind: Convection is the movement caused within a fluid (air or water) by the tendency of hotter and therefore less dense material to rise, and colder, denser material to sink, which consequently results in a transfer of heat. Rising (warm) air leaves behind a lower pressure zone on the Earth, which is back-filled by cooler air, while descending (cooler) air produces higher pressure on the surface. Air flowing from high pressure to low pressure is wind. As air rises, it cools with altitude, eventually condensing and losing its moisture as rainfall; the rising air becomes dryer.

The warm air that rises in the tropics is wet. Tropical moist air moves away from the equator and toward the poles. As it travels, it cools, dries, becomes denser, and eventually descends around 30° north and south latitudes. This dry air mass, having lost its moisture in the tropics, absorbs moisture from the ground, creating arid conditions at these latitudes. Most deserts are found within ±30° latitude.

Some of the air is drawn back toward the equator, and some is drawn toward the poles as part of a new air mass. At latitudes around ±60° the air again rises, cools and releases precipitation (though less than in the tropics). Some of the rising air then flows to the poles, where it absorbs moisture creating the cold, dry climates of the Polar Regions. Note, the Poles are very dry; in fact the South Pole is the driest continent on Earth. Precipitation amounts are very low. And some parts of the Arctic are *polar deserts* and receive about the same amount of precipitation as the Sahara Desert.

Low pressure regions occur at the equator and at ± 60° latitude where air rises; high pressure occurs at ±30° latitude and at the poles ±90° where air descends and moves towards low pressure areas. Low pressure areas are generally associated with humid climates; high pressure with dryer climates.

The process of rising and falling air, looping north and south forms three basic wind cells; *Hadley* cells, which circulate between the equator (0°) and ±30°, *Mid-latitude* or Ferrel cells that circulate between ±30° and ±60° latitude and, *Polar* cells that circulate between ±60° and ±90°. Each cell wraps around the Earth, longitudinally. The cells transport heat and moisture around the planet.

Figure 4.3-1 is a simplistic illustration of a side view of a Hadley cell; it shows warm air rising at the equator, moving toward the Polar Regions, reaching around 30° latitude, where it cools and descends to wrap around back toward the equator.

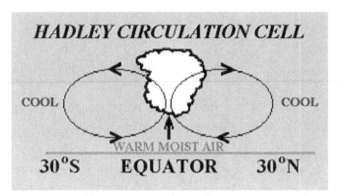

Figure 4.3-1 Illustration of Hadley Cell Circulation

Similar air movement occurs for the Mid-latitude and Polar cells as shown in Figure 4.3-2, a side view of the cell structure. This figure illustrates how variations in troposphere height, from the equator to the poles, constrain all three basic cells. (The figure shows the tropopause rather than troposphere; tropopause is at the top of the troposphere, linking the troposphere and stratosphere).

Figure 4.3-2 however, is a static view. According to Ball (2014, *The Deliberate Corruption of Climate Science*) et al, atmosphere movement is much more complicated; the troposphere height at the equator is about 10 miles, whereas the altitude of the troposphere at the Poles varies from around 5 miles in winter to around 6 miles in summer.

Dr. Ball points out the difficulties in modeling the dynamics of this air flow when the heights vary with global temperature which varies by region from day to night, season to season, and year to year.

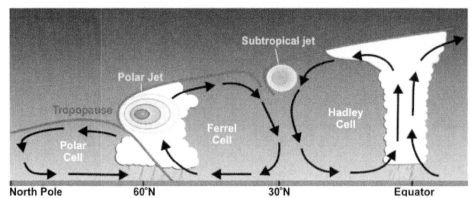

Figure 4.3-2 Side view: Troposphere (tropopause) higher at the Equator than the Poles (Source NOAA)

Figure 4.3-2 also shows that each hemisphere has two primary jet streams, a polar and a subtropical. The Polar Jet Streams form between the latitudes of 50° and 60° north and south of the equator, and the Subtropical Jet Streams are closer to the equator and take shape at latitudes close to ±30°. Jet streams are fast flowing, relatively narrow air currents found in the atmosphere around 6 miles above the surface of the Earth. They form at the boundaries of adjacent air masses with significant differences in temperature such as the polar region and the warmer air at lower latitudes.

If the Earth wasn't spinning, air flow would largely flow north and south between the equator and poles undisturbed. However, it is spinning; it rotates from west to east.

Since the Earth is rotating to the east, air has momentum in that direction. Using approximate numbers, the (nearly spherical) Earth is 24,000 miles in circumference at the equator, so the speed of land at the equator is 1,000 miles per hour (to the east). At 30° the circumference is about 21,000 miles, and the speed is about 866 mph. At 60° it's 12,000 miles, and 500 mph (half the speed of that at the equator). At the Poles, the circumference is zero and their speed of rotation is 0 mph (it's spinning on a point but no lateral movement). So, the speed of Earth's rotation at the poles is 500 mph less than that at 60° latitude, and 1,000 mph slower that at the equator. This spinning of the Earth at different speeds, varying from 1,000 mph to zero, causes the air to change direction from north to south and south to north creating those wind patterns shown in Figure 4.3-3. This is known as the *Coriolis Effect*.

Figure 4.3-3 is a representation of the Coriolis Effect on the three circulation cells. Hadley cell air rotation is responsible for *Trade Winds*; Mid-latitude cell for *Westerlies*, and; Polar cell for *Polar Easterlies*. Note, winds are labeled according to their source; e.g., southeasterly Trade Winds blow from the south-east.

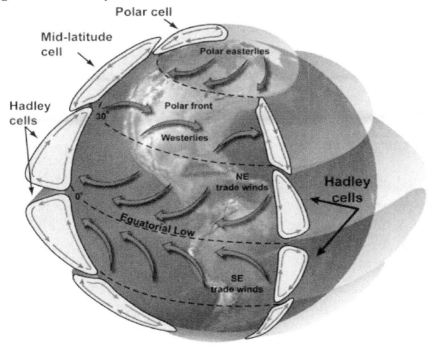

Figure 4.3-3 Air Flow and Global Wind Belts (Source: NOAA)

Because of the Coriolis Effect, circulating air is deflected toward the right in the Northern Hemisphere (N-H) and toward the left in the Southern Hemisphere (S-H). It is the inertia of the winds while the Earth spins under them that generates the appearance of winds moving east or west. Wind velocity is at maximum at the poles; i.e., Polar Easterlies at times are extremely strong winds. Though not intuitive, since there is no lateral movement at the poles, it can be seen from the equation for the Conservation of Angular Momentum that states: Mass x Radius x Velocity = Constant. As the radius reduces from 12,000 miles toward zero, mass is the same, so velocity must increase proportionally. Similarly, as the Earth's radius increases to its maximum at the equator, wind speed is at its lowest and the equator is relatively calm.

Although the latitudinal cells (i.e. Hadley, Mid-latitude, and Polar) are the major global circulations, thermal energy is also be transported longitudinally. The longitudinal circulation occurs across equatorial Pacific and is known as the *Walker cell* or *Walker circulation*. As with the three latitudinal cells, the Walker cell is also driven by temperature and pressure gradients. These winds move surface waters in a complex manner which affects global climate, particularly El Niño/La Niña.

In contrast to the Hadley, Mid-latitude and Polar circulations which run along north-south lines, the Walker cell is an east-west circulation that spans the entire width of the Pacific Ocean. Over the eastern Pacific Ocean, surface high pressure off the west coast of South America and low pressure off the eastern coats of S.E. Asia enhances the strength of easterly trade winds near the equator. Winds blow away from the high pressure at S. America toward lower pressure at S.E. Asia. Water in the eastern Pacific (west coast of S. America) is relatively cold. By contrast the water in the western Pacific (S.E. Asia) is warm (as much as 8°C warmer). The warm air over S.E. Asia (e.g., Indonesia, Australia, etc.) rises and forms clouds. This causes heavy precipitation to fall over the western tropical Pacific throughout the year. The heavy precipitation causes the rising air to dry which then circulates back at altitude towards the region in S. America near Ecuador, completing the Walker cell circulation, shown in Figure 4.3-4. The air sinks to the surface (creating high pressure) and is picked up by the strong trade winds to continue the cycle.

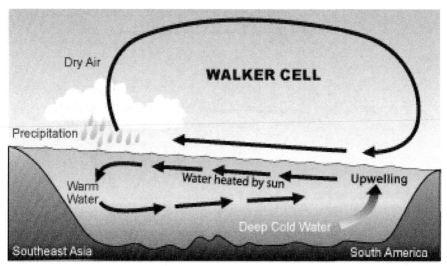

Figure 4.3-4 Walker Cell Atmospheric and Ocean Circulation

Figure 4.3-4 also shows the concomitant oceanic circulation feature associated with the Walker cell. Trade winds move surface water from the eastern Pacific to the west, where it pushes up against the coast (rising up to a foot higher than at the east coast). This water sinks (gravity) to a level where it cools, and is returned to the east Pacific coast where it upwells (rises to the surface), cooling the air.

Since the Walker cell operates along the equator (and approximately ±300 miles) it overlaps and interacts with the Hadley cell; Figure 4.3-5 illustrates the relationship around the equatorial tropical region. The Hadley and Walker circulations are the largest overturning circulations in the atmosphere covering about one third of the globe. Depending on the relative strength of the cells they either enhance or reduce the intensity of the Earth's hydrological system (water cycle), affecting the climate throughout a significant portion of the world (Asia, Africa, S. America, Australia, Indonesia, etc.).

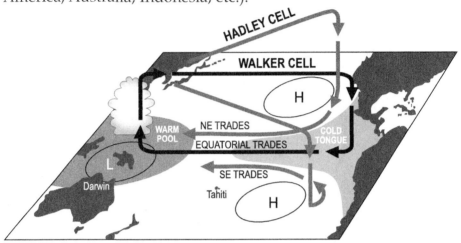

Figure 4.3-5 Interaction between the Hadley and Walker Cells over the Equatorial Tropics (Source: Adapted from NOAA Climate.gov drawing by Fiona Martin)

About every 3 to 7 years, the Walker Circulation and the trade winds weaken, allowing warmer water to reverse direction, moving from the western side of the Pacific eastward to the South America coastline. This phenomenon is known as El Niño which is discussed in section 4.4.

Winds have an enormous effect on climate and, as described above, they vary in strength, direction, altitude, latitude, etc. with varying seasons and day-to-day temperature fluctuations. Today, the Sahara is the world's largest desert. The cyclic flow of wind between the equator and sub-tropics is responsible for the existence of deserts and other arid regions (Figure 4.3-6 shows deserts within ±30° of the equator).

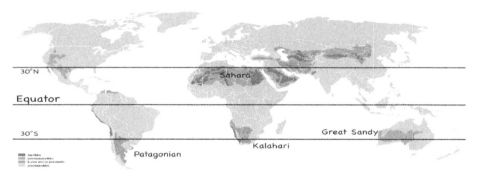

Figure 4.3-6 Deserts and Arid Regions are situated along Subtropical Latitudes

According to numerous sources (including the website phys.org/news/2016-12-years-sahara-tropicalwhat), as little as 6,000 years ago, the Sahara Desert was covered in grassland that received plenty of rainfall. And, around the same time period, according to Plimer (2009, *Heaven and Earth, Global Warming and the Missing Science*), the average sea level was more than 6 feet higher than today. Climate change without human-caused CO_2, interesting!

Ocean Currents: Oceans are a major driving force for weather and climate; they absorb and store substantially more heat than the atmosphere. Joe Bastardi (2018, *The Climate Chronicles*) estimates the capacity of the oceans to be a thousand times that of the atmosphere.

Oceans are a major component of the water and carbon cycles, discussed in section 4.5 and 4.6, respectively. Following is a brief primer:

• There are two types of ocean currents, surface and deep water. Surface currents are generated largely by wind; their patterns are a result of wind direction, the Coriolis Effect, and the position of landforms. Deep currents are produced mostly by differences in water density (temperature and salinity) and gravity; their flow patterns are a result of ocean topography - undersea mountains, ridges, valleys, plateaus, and continent edges (tectonic effects).

• Cold water is denser than warm water
• Water with high salinity is denser than water with low salinity
• Denser water sinks
• Ocean water moves toward equilibrium or balance. For example, as surface water becomes denser and sinks, the water below will rise to balance out the displaced water (upwelling)

Oceans can be divided into three basic layers. The top layer, down to around 100m (330 ft) collects the warmth and energy of sunlight. The middle or barrier layer, down to around 1,000m (3,300 ft), is called the *thermocline*; ocean temperature and density change very quickly at this layer. Below the barrier layer is the bottom layer, referred to as the *deep ocean*. It averages about 3 kilometers (2 miles) in depth. The deep ocean makes up the majority of the ocean's water volume and is much colder than the above layers. The cold ocean water is a major reservoir for absorbed CO_2, as discussed in section 4.6, the carbon cycle.

The Coriolis Effect on water at its surface is the same as for air except it affects water in *layers* down to about 100m, the *net result* of which is that currents move at 90° to surface wind direction. This is called the *Ekman Transport*. Figure 4.3-7 illustrates this phenomenon for the northern hemisphere (N-H); the southern hemisphere effects are in the opposite direction.

When surface water is moved by the force of wind, the top surface water layer drags the next layer which in turn drags the next layer, and so on. Each layer is deflected by the Coriolis Effect, with the impact lessening as the water deepens; i.e., each successively deeper layer of water moves more slowly to the right, creating a spiral effect; called the *Ekman spiral*.

The Coriolis Effect and surface friction (shear stress) cause surface currents to be directed 45° to the right of the direction of the wind. Deeper layers of water move sequentially toward the right, with each successive layer moving more slowly in layers down to about 100m. This process results in a spiral pattern

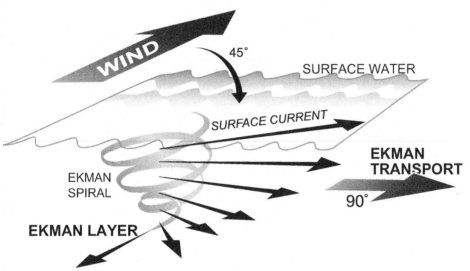

continuing down to a depth called the *Ekman Layer*.

Figure 4.3-7 Ekman Transport Effect on Current Flow

The Ekman layer is the depth at which the current is pointing in the opposite direction (180° to that of the surface wind direction) and where no further effective spiraling occurs. The speed of the wind-driven currents decreases with depth to where it becomes negligible at the Ekman layer. The *depth* of the Ekman layer is a function of the Coriolis Effect; higher latitudes (wind blowing faster) have shallower Ekman layer depths, while lower latitudes (slower wind speeds) have deeper Ekman layer depths (except for the equator where there is no Coriolis Effect).

The *Ekman Transport* is the summation of the flow (net velocity) from the surface to the Ekman layer. It is directed 90° to the right of the surface wind direction in the N-H and 90° to the left of the surface wind direction in the S-H.

Upwelling: Upwelling is a process in which deep, cold water rises to the surface to replace warm water that, because of the Ekman

Transport, is pushed 90° away from the wind direction, creating a *gap*. Colder water then rises up from beneath the surface to replace the water that was pushed away. Upwelling occurs in the open ocean and along coastlines. Figure 4.3-8 (a), top panel, shows a plan view of the prevailing surface wind and resulting current direction (north in the N-H and south in the S-H) due to the Ekman Transport. Figure 4.3-8 (b), bottom panel, shows a corresponding cross section showing the upwelling and resulting sea surface temperature anomalies, i.e., warm water shifts north and south, while cold water upwells at the equator, where it warms.

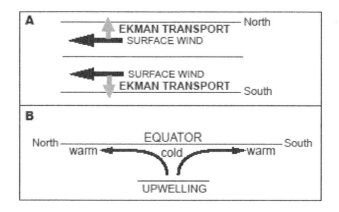

Figure 4.3-8 Ekman Transport and Upwelling

Figure 4.3-9 depicts upwelling along a continent's west coast. In the Northern Hemisphere, upwelling most commonly occurs along west coasts (eastern sides of ocean basins); e.g., the coasts of California and northwest Africa when winds blow from the north causing transport of surface water 90° away from the shore, in a westerly direction. Winds blowing from the south cause upwelling along continents eastern coasts in the Northern Hemisphere.

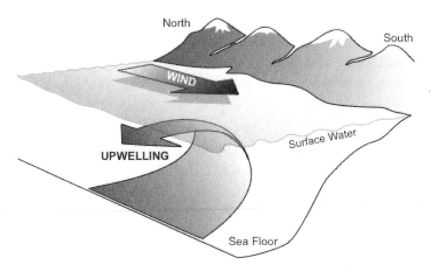

Figure 4.3-9: Upwelling along a Continent's West Coast

Surface Currents: To a large extent, horizontal movement of ocean surface waters mirrors the long-term average planetary circulation of the atmosphere. Global winds drag on the water's surface, causing it to move and build up in the directions described above.

The Westerlies of middle latitudes and the Trade Winds of the tropics drive the most prominent features of ocean surface motion, large-scale, roughly circular current systems known as *gyres*. Viewed from above, subtropical gyres rotate in a clockwise direction in the Northern Hemisphere and in a counterclockwise direction in the Southern Hemisphere.

A gyre is the circular rotation of water within an ocean basin that is driven by the wind. They are set in motion by winds and steered by the location of the continents and the Earth's rotation. The five major ocean-wide gyres are: North Atlantic, South Atlantic, North Pacific, South Pacific, and Indian Ocean gyres (Figure 4.3-10).

In the N-H, Trade Winds blow from east to west at the equator, pushing surface water to the north (Ekman Transport) to about 30° latitude where the wind shifts directions and blows from west to east (Westerlies) which together with continental barriers changes the path of the surface water to turn back down towards the southeast. This continuing pattern results in a slow clockwise rotation of water across the North Pacific Ocean and an anticlockwise rotation in the South Pacific Ocean. It can take up to 10 years for such a gyre to complete its cycle. The same phenomenon repeats itself in all five gyres around the globe with the direction of rotation depending on the hemisphere.

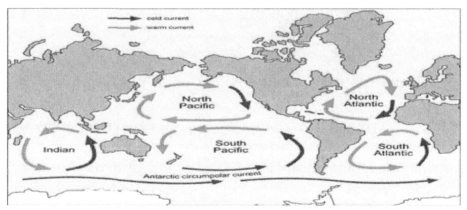

Figure 4.3-10 Major Global Gyres (Source: NOAA)

Around the world, there are some similarities in gyre currents. For example, along the west coasts of continents, currents flow toward the equator in both hemispheres, transporting colder water from Polar Regions to tropical regions. The opposite occurs along the east coasts of continents; currents move warm tropical water from the equator toward the poles. This continuous distribution of warm and cool surface water affects regional and global climates.

The South Pacific Gyre is one of the most important systems because it is very susceptible to the El Niño/La Niña effects (discussed in Section 4.4); El Niño causes massive rain storm patterns all over the world, and La Niña leads to global droughts. During either of these events the replacement of cold water by warm water causes air temperature swings and changes in humidity. This alters weather patterns by steering storms and rainfall to new locations.

The North Atlantic Gyre is also very important to global climate; it creates the Gulf Stream, keeping the countries of northwest Europe, which are at the same latitude as Greenland, relatively comfortable places to live. It also influences the climate of the east coast of the U.S. The Gulf Stream originates in the Gulf of Mexico, exits through the Strait of Florida, and follows the eastern coastline of the United States and Newfoundland. It is a powerful north-western current that exchanges heat and moisture from the water surface to the atmosphere. It has been estimated that it would take about one million nuclear power plants to provide the same energy as that provided by the Gulf Stream, without which northern Europe would have a climate similar to Siberia, i.e., about 5 to 10°C colder. Such is the power of natural climate events.

Deep currents: Ocean circulation comprises a global network of interconnected currents, counter-currents, deep-water currents, and turbulent eddies. From this complex circulation, an underlying transport pattern emerges; water cycles from surface currents to deep-water currents then back to the surface again.

Deep water circulation has a scale, pace, and power very different from surface circulation. Deep currents twist together into a continuous stream that loops through the oceans in what oceanographers refer to as the Global Conveyer Belt (GCB). The GCB results in the major oceans exchanging water with each other in a worldwide circulation cycle, which takes about 1,000 years to complete. Figure 4.3-11 illustrates the GCB.

There are two major forces driving the GCB. First there is wind. Wind, in combination with the Earth's rotation, generates the gyres that circle the major ocean basins. Turbulent swirling packets of water called eddies, many of which are hundreds of miles in diameter, spin out of these wind-driven currents and carry the water trapped inside them to other parts of the ocean.

The second force is tied to differences in water density. Temperature and salinity affect water density. The colder and saltier the water, the denser it becomes. As water becomes denser, it sinks. This density-driven component of the GCB is called the *thermohaline* circulation; *thermo* meaning heat and *haline* meaning salt.

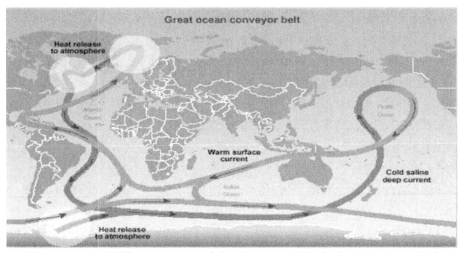

Figure 4.3-11 Illustration of the Oceans Global Conveyer Belt

(The above illustration was created by Robert Simmon, NASA. Minor modifications by Robert A. Rohde also released to the public domain - NASA Earth Observatory, (commons.wikimedia.org/w/index.php?curid=3794372)

The GCB begins on the surface of the ocean near the poles where the water is chilled by low air temperatures to freezing or below. Polar waters also get saltier because when ice forms the salt is left behind in the remaining water. As the water gets colder and saltier its density increases and it sinks toward the bottom. Surface water is pulled in to replace the sinking water which in turn becomes cool and salty enough to sink, i.e., currents begin. The GCB starts its journey at the surface of the North Atlantic, off the coast of Greenland, where great amounts of water cool and sink. Hemmed in by the continents, and its passage constrained by underwater topology this new deep water can only flow south past the equator to the far ends of Africa and S. America. Continent boundaries and ocean topology, continuously altered by tectonic activity, are significant contributors to currents flow, blocking and redirecting currents.

Starting off the Greenland coast, the newly created deep-water slowly drifts south along the western margin of the Atlantic basin. It then crosses the equator and mixes with the deep-water currents circling Antarctica.

As the current travels around the edge of Antarctica fresh streams of cold water sink into and recharge the conveyer belt. Two sections split off and turn northward, one into the Indian Ocean and the other into the Pacific. Both these currents warm up becoming less dense as they travel to the Indian and Pacific oceans, enough to where they eventually upwell to join the various gyres. Drawn by the inexorable pull of the conveyer belt, these new warm waters loop back and eventually return to the N. Atlantic to begin the 1,000-year journey again.

The ocean conveyor belt plays a crucial role in shaping the climate and it is plausible that global warming, regardless of the cause, could alter the GCB. As the Earth heats up, fresh water is formed by melting ice and increased precipitation that could dilute ocean water salinity to the point where it would not continue to sink into the depths of the ocean. (Remember, to sink, water needs to be denser, i.e., colder and saltier than the water below).

There is strong evidence that an extreme occurrence happened in the past, drastically altering the world's climate in just a matter of years. About 11,000 years ago, early in the Holocene epoch, emerging from the last glacial epoch, ice age glaciers were retreating. In central Canada, an immense glacial lake called Lake Agassiz occupied an area larger than all the Great Lakes. Suddenly the dams holding Lake Agassiz collapsed. The contents of the entire lake rushed into the North Atlantic by way of the St. Lawrence River. This massive infusion of fresh water diluted the polar seas to the point where the water was no longer dense enough to sink and the GCB most likely ground to a stop, resulting in a temporary return of a glacial period. This period, which lasted about a thousand years is called the Younger Dryas and was discussed in Chapter 3.

As an aside, while the recurrence of such an event is not considered realistic, in 2003 the Pentagon developed scenarios to survive such an incidence; it was an unclassified document printed in 2004 in Fortune magazine titled the *Pentagon's Weather Nightmare*. It describes worst-case scenarios of world reaction and possible consequences of global warming. It reads like a science fiction novel and probably inspired the movie, *The Day After Tomorrow*, a big-budget disaster film starring Dennis Quaid as a scientist trying to save the world from an ice age precipitated by global warming. No doubt this movie, though fiction, *confirmed* the beliefs of many alarmists that humans are destroying the planet.

Section 4.3 Summary: It all begins with the Sun. The Sun warms the Earth's surface more intensely at the equatorial tropics causing air to rise up toward the tropopause where it cools and circulates in various patterns (cells), north and south of the equator (Hadley, Mid-latitude, and Polar cells), and east and west (Walker cell) along the equatorial Pacific Ocean. Wind direction results from differences in atmospheric pressure and the Earth's rotation (Coriolis Effect) moving clockwise in the northern hemisphere and anti-clockwise in the southern hemisphere.

Winds move the surface water around the globe transporting heat and moisture causing such phenomena as El Niño/La Niña and the Gulf Stream.

Starting in the north Atlantic Ocean, dense water (cold and salty) sinks to the bottom of the ocean where it begins a journey known as the Global Conveyer Belt (GCB), routing its way around the globe rising in warm regions and sinking in cold regions, exchanging ocean water completely over about 1,000-year cycle. The flow of the GCB is of paramount importance to the equilibrium of the planet's climate, and a change in the flow of the GCB could have a substantial impact.

4.4: Oceanic Oscillations:

There are numerous oscillations in the Pacific, Atlantic, Antarctic, Arctic, and Indian oceans that contribute to the complexity of climate. In his discussion of ocean oscillations effect on climate, meteorologist Joe Bastardi (2018, *The Climate Chronicles*) says,

"... simple oceanic climate cycle theory explains perfectly what is going on."

This section describes four major oceanic oscillations; El Niño/La Niña, PDO, AMO, and NAO. While each of these oscillations has enormous impact on the Earth's climate, their causes are poorly understood and, as such, cannot be predicted by climate models.

El Niño/La Niña: El Niño and La Niña are opposite phases of a natural climate pattern across the tropical Pacific Ocean that swings back and forth every 3-7 years on average. Together, they are called ENSO, which is an acronym for **El Niño-Southern Oscillation.** ENSO is a natural phenomenon arising from coupled interactions between the atmosphere and the ocean in the tropical Pacific.

ENSO can be in one of three states: Neutral, El Niño, or La Niña. Neutral indicates conditions that are near their long-term average in terms of atmospheric and oceanic conditions: air pressure and circulation, trade wind speed and direction, oceanic surface and sub-surface temperature and direction, and the consequential effects on regional and global climate. El Niño, the warm phase, and La Niña, the cool phase, lead to significant climatic differences from the neutral state.

Neutral: Neutral describes the *average* state of the atmospheric and oceanic circulation patterns over the tropical Pacific. Normally, sea surface temperature (SST) is about 8°C higher in the western Pacific than the water in the eastern Pacific. Neutral conditions are as described in section 4.3 and illustrated in Figure 4.3-4. Restating the neutral atmospheric and oceanic conditions:

Atmospheric Circulation:
• The eastern side of equatorial Pacific (South America) is a high-pressure zone – cool air descends
• The western side (S.E. Asia) is a low-pressure zone – warm air rises
• Trade winds blow from the east Pacific to the west Pacific, along the equator
• On the western side of the Pacific, warm air rises, cools and condenses into heavy rainfall; losing its moisture, it dries with altitude
• Upon reaching the troposphere, it travels eastward from S.E. Asia toward S. America (remember there is no Coriolis Effect along the equator)
• When it reaches the west coast of S. America, the cool air descends, completing the *Walker Cell* circulation pattern of air flow

Oceanic Circulation:
• Trade winds move warm surface water from east to west (S. America to S.E. Asia)

- Upon reaching western Pacific land, the blown water causes the sea level to rise (about one foot) higher than at the eastern side of the Pacific
- This, in turn, causes the warm water to descend (gravity) where, at the thermocline, it cools. (thermocline is the transition layer between the warmer mixed water at the surface and the cooler deep water below at which there is a sudden change in temperature)
- The cool/cold water circulates eastward toward S. America
- Upon reaching the S. America coastline, the cool/cold water upwells to balance the higher sea level at the eastern coast of the Pacific (at S.E. Asia). This completes the oceanic circulation path

During a neutral ENSO, the western side of the Pacific (S.E. Asia – Indonesia, Philippines, Australia, etc.) experiences heavy rainfall, monsoons, etc.; the eastern side (S. America, S.W. USA, etc.) encounters dry conditions, often resulting in droughts.

El Niño: El Niño, the warm phase of ENSO, occurs when air pressure patterns *reverse* such that the eastern Pacific is low pressure and the western Pacific is high pressure. Trade winds weaken, stop, or even reverse direction such that atmospheric and oceanic flow directions are opposite to those shown in Figure 4.3-4. The atmospheric Walker Cell flows counter-clockwise (side view) while the ocean flow is clockwise, upwelling cool water along the coasts of S.E. Asia. This results in a warming and increased moisture of the western Pacific and cooling and drying climate over the eastern Pacific.

The Walker cell, which extends to the Indian and Atlantic Oceans, interacts with the Hadley cells (discussed in section 4.3) over the equatorial Pacific to impact global atmospheric circulation. By this interaction, the rising air in the tropical Pacific can branch away from the equator toward the higher latitudes, both northward and southward, contributing to the complex circulation patterns that help establish the average worldwide climate. When an El Niño causes excess heating in the tropical Pacific upper atmosphere, air flow toward the poles becomes more vigorous. Changes in the strength of the Hadley circulation leads to further changes in atmospheric circulation patterns worldwide.

During El Niño, higher-than-normal surface air pressures develop over Australia, Indonesia, Southeast Asia and the Philippines, producing drier conditions or even droughts. Dry conditions also prevail in Hawaii, parts of Africa, and northeastern Brazil and Colombia. The eastern side of the Pacific receives heavy rainfall. There is an inverse correlation between intense El Niño events and hurricane activity; i.e., in *very strong* El Niño years there are fewer hurricanes.

According to Plimer (2009, *Heaven and Earth*), despite El Niño being one of the greatest transfers of energy on Earth, they cannot be predicted by computer models and are not featured in models of future climate.

La Niña: La Niña, the cool phase of ENSO, which often follows an El Niño, can be best described as "ENSO-neutral on steroids." The general states of the neutral ENSO remain, but the elements that comprise the neutral state, atmospheric and oceanic circulations, are intensified. The pressure gradient between the eastern Pacific (high) and the western Pacific (low) is greater than neutral, and trade winds are stronger. This leads to higher sea levels at S.E. Asia and an increased upwelling of colder water off South America.

Eastern to Central Pacific Ocean water and air temperatures cool and the climate tends to be drier than when neutral. At the western side of the Pacific, S.E. Asia experiences heavier rainfall, monsoons, etc. Regions that are typically dry during El Niño tend to become excessively wet during La Niña, and vice versa.

The Oceanic Niño Index (ONI) is the de-facto standard used by NOAA for identifying the intensity of El Niño and La Niña events. ONI levels for El Niño vary from *weak* to *very strong*; for La Niña, it is measured from *weak* to *strong*. Figure 4.4-1 shows maps of sea surface temperature (SST) anomaly in the Pacific Ocean during *strong* La Niña (top – dark area, cold SST) that occurred in 1988 and the 1997 *very strong* El Niño (bottom – dark area, warm SST).

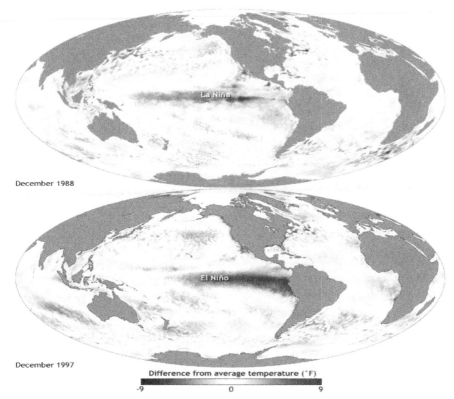

Figure 4.4-1 Maps of sea surface temperature anomaly in the Pacific Ocean during strong La Niña (top, December 1988) and El Niño (bottom, December 1997). Source: NOAA Climate.gov)

Figure 4.4-2, provided by the National Center for Atmospheric Research (NCAR), shows El Niño (top) and La Niña (bottom) phases of abnormal sea surface temperatures from 1950 to 2017.

The ONI shows three *very strong* El Niño's: 1982/83, 1997/98, and 2015/16. Each of those El Niño's correlated with significant increases in global temperature of around 0.6°C. NOAA suggests that the dominance of La Niña's from 1998 to 2015 may have been responsible for *the pause* of global temperature over that period. The IPCC however, suggests that *aerosols* were probably the reason. They just don't know!

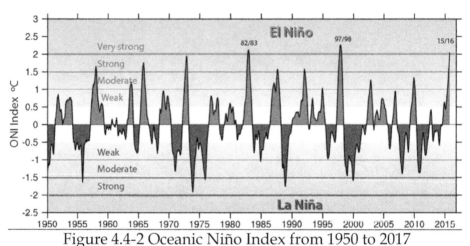

Figure 4.4-2 Oceanic Niño Index from 1950 to 2017
(Source: NCAR: Author Dr Kevin Trenberth, Distinguished Senior Scientist)

In addition to affecting global atmospheric circulations, which move heat and moisture globally, there is evidence that El Niño (1) increases atmospheric CO_2 and, (2) affects cloud cover.

(1). Atmospheric CO_2: in 2014, NASA launched a satellite whose mission was to examine the effects of the *very strong* 2015/2016 El Niño on atmospheric CO_2. Dr. A. Chatterjee, a scientist at the Global Modeling and Assimilation Office, et al, published the results of the mission (2017, *Science, Vol 358*), which showed a net increase in atmospheric CO_2 concentrations attributed to the El Niño event. Much of the increase was apparently from forest fires in drought regions.

(2). Clouds: A 2013 article by Roy W. Spencer and William D. Braswell (*Asia-Pacific Journal of Atmospheric Sciences*) describing climate modeling research they performed, says,

"A natural shift to stronger warm El Niño events in the Pacific Ocean might be responsible for a substantial portion of the global warming recorded during the past 50 years."

Spencer and Braswell included changes in cloud cover, consistent with satellite observations, in their modeling. The results suggest that these natural climate cycles, El Niño/La Niña, change the amount of energy received at the Earth' surface by virtue of increasing or decreasing cloud cover. They suggest that as much as **50% of the warming since the 1970s could be attributed to strong El Niño** activity.

Pacific Decadal Oscillation (PDO): PDO is a pattern of climate variability similar to ENSO in character but which varies over a much longer time scale and mostly affects *northern* Pacific temperatures rather than the tropical Pacific. The PDO refers to cyclical variations in sea surface temperatures (SST) in the Pacific Ocean; each phase, warm and cool, occurs every 20-30 years, so that a complete PDO cycle lasts around 40-60 years. The PDO warm and cool phases, extending from the Equator northward along the coast of North America into the Gulf of Alaska, alter atmospheric and oceanic circulation patterns. Shifts in a PDO phase can have powerful implications for global climate, affecting hurricane *intensity*, droughts and flooding around the Pacific basin, and global land temperatures. Note, the Pacific Ocean occupies about 40% of the Earth's surface so it is reasonable to assume that changes in its composition affect global climate.

PDOs can modulate the impacts of ENSO according to its phase. If they are in the same phase, El Niño/La Niña impacts are magnified. Conversely, if they are out of phase, they offset one another, preventing *true* ENSO impacts from occurring.

A blog by Dr. Roy Spencer shows the cold and warm cycle phases of the PDO from 1900 to 2008 (Figure 4.4-3). The PDO was in a warm phase from about 1915 to about 1945, a cool phase from about 1945 to 1977, warm from 1977 to 1998, and cool beginning in 1999, demonstrating remarkable synchronicity with average global temperature records. But because of its cycle period (40 to 60 years), and satellite observations have only been available since the 1970s, there is insufficient sample size to draw firm conclusions.   (drroyspencer.com/global-warming-background-articles/the-pacific-decadal-oscillation).

Global temperatures are tied directly to SST. When SSTs are cool (cool phase PDO) the global climate cools. When SSTs are warm (warm phase PDO), the global climate warms. Specifically, note that in 1945 the PDO switched from its warm mode to its cool mode and global climate cooled from then until 1977 (Remember media warnings of the next *Ice Age*). In 1977, the PDO switched abruptly from its cool mode into its warm mode and global climate shifted from cool to warm. This rapid switch from cool to warm has become known as *The Great Climatic Shift*; it signaled the beginning of what became known as *global warming*. Since the PDO decline starting in the late 1990s global temperature stabilized through around 2014. According to the *Joint Institute for the Study of the Atmosphere and the Ocean*, causes for the PDO are not currently known.

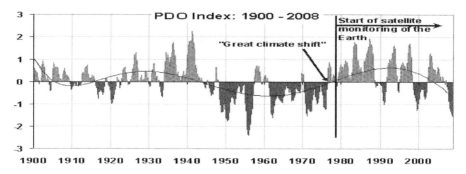

Figure 4.4-3 PDO Index from 1900-2008

Dr. Spencer (2010, *The Great Global Warming Blunder*) suggests that the PDO is critical to our understanding of global warming because **a change in circulation patterns can cause a change in global-average cloudiness and, he says, a change in cloudiness associated with the PDO might explain most of the climate change seen in the last 100 years, including 75% of the warming trend.** As discussed elsewhere in this book, clouds are the single largest internal factor effecting global temperatures and, they represent the largest source of uncertainty in IPCC models.

Satellite monitoring of Arctic sea ice began in 1979 when the PDO went from its negative to positive phase. The satellites showed a warming Arctic region with gradual loss of ice cover during the yearly summer melt season. In late 2007, the Northwest Passage opened, a rare event, which allows ships to travel the relatively short distance between the Atlantic and Pacific Oceans through the Arctic.

But while the media alarms the public marking this as catastrophic and attributing it to recent increase in anthropogenic CO_2, they fail to note that similar events occurred in the 1930s, with disappearing sea ice and the opening of the Northwest Passage in 1939 and 1940. Arctic temperatures were just as warm then as they are now when atmospheric CO_2 was around 300 ppm, well below even the most aggressive IPCC et al targets.

Atlantic Multidecadal Oscillation (AMO): AMO a climate cycle of long-duration changes in the sea surface temperature of the North Atlantic Ocean, with cool and warm phases that last for 20-40 years. These changes are natural and unpredictable. Instruments have observed AMO cycles only for the last 150 years, and satellite observations only for the past 40 years. However, according to NOAA, studies of paleoclimate proxies have shown that oscillations similar to those observed instrumentally have been occurring for at least the last millennium.

Research suggests that the AMO is related to the past occurrence of major droughts in the U.S. Midwest and the Southwest. When the AMO is in its warm phase, these droughts tend to be more frequent or prolonged. Two of the most severe droughts of the 20th century occurred during the positive AMO phase between 1925 and 1965.

AMO affects not only the Northern Hemisphere but global climate. Through changes in atmospheric circulation (wind and clouds), AMO can alter spring snowfall over the Alps, rainfall patterns in North Eastern Brazil and African Sahel. The African Sahel is a semi-arid region of western and north-central Africa extending from Senegal eastward to the Sudan to the river Nile. It forms a transitional zone between the arid Sahara Desert to the north and the belt of humid savannas to the south. And climate models, confirmed by Paleo-climatologic studies, suggest that a warm phase of the AMO strengthens the summer rainfall over India.

Although the cause of AMOs is not fully known, in 2010 a Norwegian research team demonstrated that the phases of the AMO over the past 600 years were controlled mainly by fluctuations in solar activity and large volcanic eruptions. Other studies, including one by Kuhnert and Mulitza (2011, *Paleoceanography* 26 (4), PA4224), also found a solar influence on AMOs.

North Atlantic Oscillation (NAO): Is a weather phenomenon in the North Atlantic Ocean of fluctuations in the difference of atmospheric pressure at sea level between the Icelandic low and the Azores high; the Azores are located west of Portugal in the Atlantic. NAO is a natural, periodic change (positive and negative phases), in atmospheric pressure between the latitudes of Iceland (65°N) and the Azores (40°N) – about 1,500 miles - that affect the strength of prevailing Westerlies over the North Atlantic Ocean. According to Professor Richard J. Greatbatch of the Department of Oceanography, Dalhousie University (May 4, 2000, *The North Atlantic Oscillation*), NAO is *"the most important mode of atmospheric circulation variability in the northern hemisphere."* NOA plays a major role in weather and climate variations over Eastern North America, the North Atlantic and the Eurasian continent.

NAO exerts a dominant influence on *wintertime* temperatures across much of the Northern Hemisphere with durations varying widely from short-term fluctuations of 2 to 5 years, with superimposed oscillations of 12 to 15 years, and also about 70 years. The pressure difference between Iceland and the Azores controls the strength and direction of westerly winds and strength and location of surface currents and storm tracks across the North Atlantic.

During its *positive* phase, strong high pressure over the Azores, strong low pressure over Iceland (maximum pressure gradient), NAOs are associated with a more powerful Gulf Steam and above-normal temperatures and precipitation in the eastern United States and northern Europe, and below-normal temperatures and precipitation over central and southern Europe. In its *negative* phase, the pressure over Iceland and Azores are similar, a weak low pressure and weak high pressure over Iceland and the Azores, respectively (minimum pressure gradient). This weaker pressure gradient causes winter storms to track further south bringing unsettled weather into the Mediterranean and southern Europe while colder, drier air affects northern Europe.

James W. Hurrell and Harry van Loon of the National Center for Atmospheric Research (July 1997, Volume 36, *Decadal Variations in Climate Associated with the North Atlantic Oscillation*), say that from 1980 through 1997, a period of intense average global warming, NAO accounted for a substantial part of the observed wintertime surface warming over Europe and downstream over Eurasia, and cooling in the northwest Atlantic, and that estimates based on 60 years of data from 1935 to 1995 show that the **NAO accounted for 31% of the variance in hemispheric winter surface air temperature north of 20°N**. These findings are significant since global temperatures reported by the IPCC et al are averaged over an entire year including all seasons and both hemispheres.

Section.4.4 Summary: When ocean and atmospheric conditions in one part of the world change as a result of ENSO (El Niño /La Niña) or any other oscillation (PDO, AMO, NAO, et al), the effects are felt around the world. The rearrangement of atmospheric pressure, which governs wind patterns and sea-surface temperature, can drastically affect regional and, in many cases, global weather patterns. Since their causes are unknown, their effects on climate cannot be effectively modeled. And their effects on global climate can be significant; e.g., the 2015/2016 El Niño is considered to be responsible for the northern hemisphere experiencing the highest temperature since the 1930s.

Within any given decade, the warmest years are usually El Niño ones, and the coldest are usually La Niña ones. That's because the Pacific Ocean occupies more than 40% of the planet and ENSO events change atmospheric and oceanic circulations and intensities over the equatorial Pacific and also the atmospheric circulation from the equator north and south interacting with the Hadley cells.

The PDO shows remarkable synchronicity with global temperature variations. Its cold phase closely tracks the *"coming of the next Ice Age"* period from around 1945 to 1977; its warm phase correlates well with average global warming from late 1970s through 1998 when the PDO switched to a warm phase. However, because of its 40 to 60-year cycle period, and satellite observations have only been available since the 1970s, there is insufficient sample size to draw firm conclusions.

Research suggests that the AMO is related to past occurrences of major droughts in the U.S. Midwest and Southwest. When the AMO is in its warm phase, these droughts tend to be more frequent or prolonged. Two of the most severe droughts of the 20th century occurred during the 40-year positive AMO phase between 1925 and 1965. There is evidence that solar activity influences AMOs.

NAO affects the strength of prevailing Trade Winds and the Westerlies, over the North Atlantic Ocean, and according to Professor Richard J. Greatbatch the NAO is the most important mode of atmospheric circulation variability in the northern hemisphere. NOA plays a major role in weather and climate variations over eastern North America, the North Atlantic and the Eurasian continent.

4.5 Clouds and Aerosols:

Two of the most important and least understood elements in climate studies are the effects of clouds and aerosols; they are the Achilles heel for climate modeling and forecasting. The IPCC admits, in its own assessment reports (AR), that it does not clearly understand how to include their effects in its models. Following are excerpts from the Third (2001), Fourth (2007), and Fifth (2013) IPCC ARs (Working Group I scientific papers, NOT the politicized Summary Reports):

Third IPCC report (TAR):

"The single greatest source of uncertainty in the estimates of the climate sensitivity continues to be clouds...."

This statement should paralyze AGW alarmists – climate sensitivity, i.e., the effects on climate of doubling CO_2 levels, is the basis for their model forecasts and Armageddon alarmism.

"...we find that, there has been no apparent narrowing of the uncertainty range associated with cloud feedbacks in current climate change simulations.... Reducing the uncertainty in cloud-climate feedbacks is one of the toughest challenges facing atmospheric physicists...."

Understanding feedback is paramount to understanding the behavior of any system. If you don't understand the feedback processes of a system, you don't understand how the system functions.

"Cloud modeling is a particularly challenging scientific problem· ...." See Figure 4.5-5 later in this section to see how understated this comment is.

"Unless there are stronger links between those making observations and those using climate models, then there is little chance of a reduction in the uncertainty in cloud feedback in the next twenty years."

This statement says that the models don't work - *"If it disagrees with observations – it's wrong. That's all there is to it."*

Six years later, from the Fourth IPCC report (AR4):

".... models exhibit a large range of global cloud feedbacks, with roughly half of the climate models predicting a more negative CRF (Cloud

Radiative Forcing) in response to global warming, **and half predicting the opposite."**

Half predicting more *negative* CRF, and half predicting more *positive*? Not very compelling – the billion-dollar computer system could apparently be replaced by a coin!

And from the Fifth IPCC AR5, in which clouds and aerosols were combined:

"... the challenges remain daunting.... the representation of both clouds and aerosol–cloud interactions in large-scale models remains primitive."

After 30 years investing billions of dollars in research and technological advancement, the problems are still *daunting* and the models remain *primitive!*

Clouds and aerosols continue to contribute the largest uncertainty to estimates and interpretations of the Earth's changing energy budget...."

This was discussed in section 4.1, and it is clear that the reported CO_2 contribution to the energy budget is noise level compared to that of clouds and aerosols.

4.5-1 Clouds: *"Whatever controls the clouds, rules the climate."* (Vahrenholt & Luning, *The Neglected Sun*, 2015).

According to Voilard (2010, *NASA Earth Observatory*),

"Just a 5% increase in cloud reflectivity could compensate for the entire increase in greenhouse gases from the modern industrial era in the global average."

Plimer (2009) suggests that a *"1% change in cloudiness could account for all the 20th century warming."*

And Dr. John Holdren, President Obama's science advisor for eight years (1972, *Global Ecology: Readings Toward a Rational Strategy for Man*) said,

" ...a mere one percent increase in cloud cover would decrease the surface temperature by 0.8°C."

Note, 0.8°C is the upper level of the generally agreed temperature increase over the entire 20th and 21st centuries.

Considering the IPCC's expressed uncertainty in their understanding and modeling of clouds, this alone should introduce doubt in the anthropogenic CO_2 position.

A NASA study of clouds, where they occur, and their characteristics shows that they play a key role in understanding climate change (Earthobservatory.nasa.gov/Features/Clouds). And, according to Dr. Gavin Schmidt, Director of NASA's Goddard Institute for Space Studies (GISS) (2010, *Taking the Measure of the Greenhouse Effect*), **clouds contribute more to the greenhouse effect than CO_2,** 25% and 20%, respectively – and that's total CO_2, of which about 5% is human-caused, so about 1% is attributable to humans, or clouds contribute 25x more to the climate than human-caused CO_2.

Low, thick clouds primarily reflect solar radiation and cool the surface of the Earth. High, thin clouds are primarily transparent to incoming solar radiation; at the same time, they trap some of the outgoing infrared radiation emitted by the Earth thereby warming the surface. Whether a given cloud heats or cools the surface depends on several factors, including the properties of the aerosol particles that form the cloud, the cloud's altitude, temperature, size, opacity, and density, all of which continuously vary as they interact with other climate elements, the Sun, wind, etc.

Clouds have an almost infinite number of property configurations. Clouds affect the climate but changes in the climate, in turn, affect clouds. This relationship creates a complicated system of climate feedback (that IPCC scientists admit they do not understand) in which clouds modulate the Earth's radiation and water balances.

According to the *World Meteorological Organization's International Cloud Atlas*, over 100 types of clouds exist. Each of these can be divided into one of ten basic types depending on its general shape and altitude level: low, middle, and high. The ten types are:

• Low level clouds (*cumulus, stratus, stratocumulus*) below 6,500 feet
• Middle clouds (*altocumulus, nimbostratus, altostratus*) between 6,500 and 20,000 feet
• High level clouds (*cirrus, cirrocumulus, cirrostratus*) above 20,000 feet
• Then there is *cumulonimbus* which towers across the low, middle, and upper levels

High level clouds are made of tiny ice crystals. Mid-level clouds are usually formed from water droplets but can also include ice crystals at the top of the cloud if the temperatures are cold enough. Lower-level clouds are formed from water droplets. These ten can be further reduced to the following four types based on appearance or shape: cumulus (fluffy with dome shape); stratus (flat, thin); cirrus (wispy), and the fourth; nimbus (dark grey) which are associated with rain and/or thunderstorms.

Clouds form by several means: first, by air convection, i.e., air rises, cools, condenses. The following processes cause air to rise:

- Convection of air heated by the Sun
- Frontal forcing of cold air converging with warmer air, which is pushed upwards
- Orographic, or topological, such as when air hits a mountain and rises
- Convergence of warm air from different directions causes resultant air to rise

As air rises, it expands with decreasing atmospheric pressure; the molecules spread apart using up heat energy that results in cooler air. As it rises, it cools until its relative humidity reaches the point of saturation (100% relative humidity – dew point) when it condenses into liquid form. At this point it combines with aerosol particles to form clouds.

Aerosols are minute solid and liquid matter such as smoke particles from fires or ash from volcanoes, ocean spray, or microscopic specks of wind-blown soil. Called cloud condensation nuclei, these particles are about $1/100^{th}$ the size of a cloud droplet upon which water condenses. All cloud droplets have a speck of dirt, dust, or salt crystal at their core. Without aerosol particles there would be no clouds.

Aerosols are suspended unevenly throughout the atmosphere, and this uneven distribution affects cloud formation. The IPCC readily acknowledges that modeling clouds is the biggest source of uncertainty. One reason for example, is that according to Spencer (2009, *Climate Confusion*), two clouds having identical altitudes, thickness, and water content can have very different effects on the climate. This is because the size of the cloud *droplet* (that make up the clouds), resulting from the properties of the particular aerosol(s) has a huge impact on its reflection ability (albedo) and precipitation. Fewer aerosols (cleaner air), result in more droplets per aerosol particle such that each droplet contains more water; this produces dark clouds with heavy droplets that precipitate (rain) and therefore are shorter lasting and have relatively low albedo. More aerosols (higher polluted air), have the opposite effect; they have fewer water droplets per aerosol particle resulting in larger, brighter more reflective clouds (high albedo) that are longer lasting since they don't produce rain. Figure 4.5-1 illustrates this phenomenon.

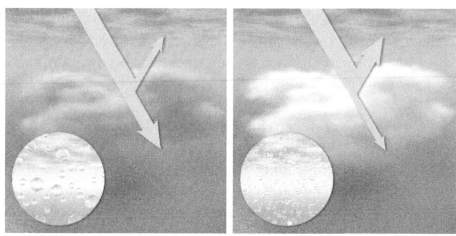

Figure 4.5-1 (On previous page) *"Clouds in clean air are composed of a relatively small number of large droplets (left). As a consequence, the clouds are somewhat dark and translucent. In air with high concentrations of aerosols, water can easily condense on the particles, creating a large number of small droplets (right). These clouds are dense, very reflective, and bright white.* **This influence of aerosols on clouds... is a large source of uncertainty in projections of climate change.** *"*(NASA image by Robert Simmon)

However, according to a NASA article by Han et al. (1998, *Global survey of the relationships of cloud albedo and liquid water path with droplet size*), it's not so straight forward; the expected decrease in cloud droplet size associated with larger aerosol concentrations has been found to be larger over land than over water and larger in the Northern than in the Southern Hemisphere, but the corresponding cloud albedo effect has not been found. They found that cloud albedo increases with decreasing droplet size for most clouds over continental areas and for all optically thicker clouds (as described above), but that cloud albedo decreases with decreasing droplet size for optically thinner clouds over most oceans and the tropical rain forest regions. This is a mere microcosm of problems in understanding cloud behavior and, while it is not surprising that cloud behavior creates such uncertainty for IPCC scientists, it *is* surprising that the IPCC maintains its 95% confidence in its models. And it is more surprising (or perhaps not, given that it's about politics, not science) that climate change debates almost never include discussions on cloud behavior.

Aerosol *type* plays an important role in determining how aerosols affect clouds. Whereas reflective aerosols, sand, dust, etc., tend to brighten clouds and make them last longer, black carbon from soot has the opposite effect. And there is a continuum of cloud density based on the distribution, size, and characteristics of specific aerosols. Forming clouds is considered an *indirect* effect of aerosols on the climate; aerosols have *direct* effects on climate, which are considerable, and discussed in the next section.

Another way clouds form is when warm air passes over a cold surface to produce fog; this is called *advection fog*. And finally, as discussed in the earlier section on the Sun, much research shows evidence of clouds being formed by galactic cosmic rays entering the atmosphere, with the degree of cloud formation being a function of the solar magnetic field: stronger fields intercept rays and reduce cloud formation and vice versa.

The atmosphere is in constant motion: as air rises, drier air is added into the rising parcel so that both condensation and evaporation are continually occurring. A further complication in modeling the effects of clouds on climate is that generally clouds do not last long in the state that they are first formed; they are constantly evolving and evaporating (or melting) depending on their altitude and particle composition. Storm clouds (cumulonimbus) can persist for an hour to over a day depending on the strength of the storm that is producing them. Fair weather cumulus clouds usually have a life span of 5 to 40 minutes. There are several types of clouds that can last on the order of a day given the right atmospheric conditions. It depends on the type of cloud and the state of the atmosphere. On average, cloud life is several hours.

Just as the altitude of the wind cells change with the varying height of the troposphere (Figure 4.3-2), cloud height is limited by altitude of the troposphere which decreases from its maximum at the equator (about 10 miles average between winter and summer) to a minimum at the poles (between 5 and 6 miles). This phenomenon, illustrated in Figure 4.5-2, which is continuously changing, adds to the complexity of cloud modelling.

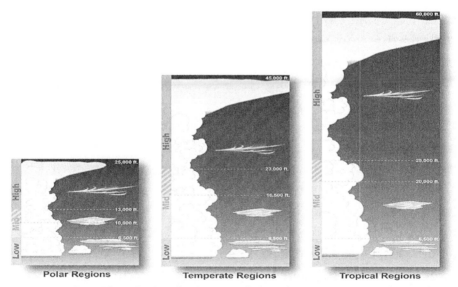

Figure 4.5-2 Cloud Height Varies from the Equator to the Poles
(Source: NOAA; National Weather Service)

Clouds modulate the Earth's radiation balance both in the shortwave (reflecting and absorbing visible light) and long wave (absorbing and re-emitting infrared) ranges, and clouds are also an essential ingredient in the hydrological or water cycle. Both of these effects are discussed below.

Radiation Effect of Clouds: The Earth's climate system constantly adjusts in a way that tends toward maintaining a balance between the energy that reaches the Earth from the Sun, and the energy that is returned into space from the Earth. Whether clouds absorb or reflect radiation depends on specific cloud characteristics; following are the two extreme effects caused by high clouds and low clouds:

High Clouds: Thin cirrus clouds act in a way similar to clean air in that they are highly transparent to the Sun's shortwave radiation (their cloud albedo or reflectivity is small) entering the atmosphere, but they absorb outgoing long wave (infrared) radiation being emitted from the Earth. Because cirrus clouds are high in the atmosphere (above 20,000 ft) they are cold (ice crystals), so that the long wave radiation is mostly retained in the clouds; i.e., a net warming.

Low Clouds: In contrast to the warming effect of the higher clouds, low cumulus clouds cool the Earth. Because low clouds are much thicker than high cirrus clouds, they are not as transparent: they do not allow as much solar energy to reach the surface. Instead, they reflect much of the sunlight back into space; the net effect of these clouds is cooling.

But of course, these are stereotypical and static examples of cloud behavior. Clouds do not fit neatly into clean categories: remember there are more than 100 types; they appear and vanish; their altitude, temperature, size, opacity, and density change, and; their properties are complex (including the intricate characteristics of aerosols) and vary in real time.

Hydrologic effect of Clouds: this involves the transfer of water and heat from the oceans to the land surfaces and back to the oceans. Everything starts with the Sun: water evaporates, condenses, and forms clouds which are moved around the globe by atmospheric circulation releasing water back to the Earth by precipitation. Clouds can be carried along by winds of up to 150 mph or can remain stationary while the wind passes through them. Figure 4.5-3 illustrates the complexity of the hydrologic cycle that is governed principally by the Sun and clouds.

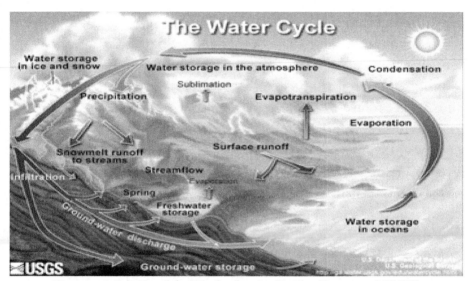

Figure 4.5-3 Earth's Hydrologic Cycle (Source: USGS)

There are four main stages in the water cycle: evaporation, condensation, precipitation, and collection. The hydrologic cycle begins with the evaporation of water primarily from the surface of the ocean and through evapotranspiration (the process by which water is transferred from land to the atmosphere by evaporation from soil and other surfaces and by transpiration from plants and trees). As moist air is lifted, it cools and water vapor condenses to form clouds. Moisture is transported around the globe by winds until it returns to the surface as precipitation. Once the water reaches the ground, one of two processes may occur: 1) some of the water evaporates back into the atmosphere forming new clouds, or; 2) it penetrates the surface and become groundwater.

Groundwater either seeps its way to into the oceans, rivers, and streams, or is released back into the atmosphere through transpiration of vegetation, again creating clouds. The balance of water that remains on the Earth's surface is runoff, which empties into lakes, rivers and streams and is carried back to the oceans, where the cycle begins again.

In a complicated feedback process clouds and climate are tightly integrated: when the climate changes clouds change; when clouds change the climate is impacted. What is important is the sum of the competing effects, the net radiative and water cycle cooling or warming effect of all clouds in the atmosphere over time. But it is not known precisely what types of cloud characteristics/interactions diminish warming (negative feedback) or enhances warming (positive feedback). Clearly both types of feedback occur; it is the degree of each that is in question and a major source of disagreement, and therefore uncertainty, among scientists.

Whereas the IPCC considers mostly positive feedback in this context, Lindzen et al (2001, 2009, 2013) present evidence of a phenomenon called the Iris Effect, whereby tropical cirrus clouds decrease with increasing sea surface temperature (evaporation and precipitation occurs) allowing heat to pass into space, presenting a cooling mechanism (negative feedback). The Iris Effect is discussed in section 4.6.

On average, according to NASA, clouds cover about 67% of the Earth at any time, i.e., they dominate the sky. Figure 4.5-4 is an average of 13 years of cloud cover data, from July 2002 to April 2015, collected by NASA's Aqua satellite. And remember, according to President Obama's Science Czar, an almost imperceptible change in cloudiness (1%) could account for the entire warming (0.8°C) since the late 19th century. Can anyone really distinguish between cloud coverage at 67% and 68%?

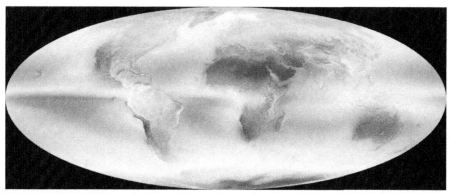

Figure 4.5-4 "The darker the color, the less cloudy it is. It's easy to see where all the deserts are. Africa is dominated by darker hues, even beyond the Sahara, as well as the Middle East and Australia. In the Americas, the U.S. southwestern states, including drought-stricken California, stand out with their darker blues, while in South America, the near-zero cloud cover of Chile's Atacama Desert is a stark contrast to the whiter hues of the rest of the continent." (Source: NASA).

Given the immense uncertainty associated with cloud simulation, it is not surprising that the climate models used by the IPCC are not very good at replicating cloudiness as can be seen from Figure 4.5-5. The darkest line represents the observed values and the other lines represent the various climate models used by IPCC. Clearly, no useful information can be garnered from these data.

The IPCC Working Group report, (IPPC Report, AR5, 7.1.2.), admits its uncertainty,

"The representation of cloud processes in climate models has been recognized for decades as a dominant source of uncertainty in our understanding of changes in the climate system." Yet, it maintains its 95% confidence in its forecasts!

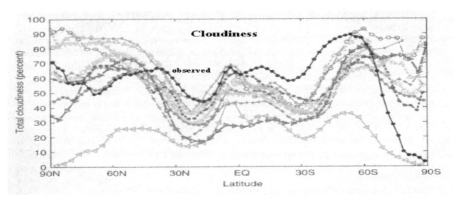

Figure 4.5-5 Cloudiness during a Northern Hemispherical Winter (Source: IPCC, Third Assessment Report: *Climate Change*, 2001)

Finally, consider that on a summer day with no clouds the temperature is noticeably warmer; with clouds, it is cooler. Without cloud cover, the temperature drops sharply at night whereas with clouds the temperature drop is noticeably more moderate. Since the greenhouse effect from carbon dioxide in a particular region would be the same on a clear or a cloudy night it could be inferred, anecdotally, that **the effect from CO_2 is negligible compared to that of clouds**.

Clouds are formed by the actions of the wind, oceanic circulations, Sun directly (through evaporation), and indirectly (cosmic rays); there are more than 100 types of clouds; they have an infinite number of properties; they are too small to be captured by models (discussed in Chapter 5), and their existence can be ephemeral. Along with aerosols, they are most difficult element to understand and predict their behavior.

4.5-2 Aerosols: While aerosols are an essential component of cloud formation, they are in their own right a key ingredient in affecting the climate. I believe that most of the general public, including me until I did this research, think of aerosols in the context of hair or paint spray, refrigerants or fire suppressants, and other products that were found to damage the Ozone layer and were addressed by the Montreal Protocol. However, these microscopic particles exist everywhere in the air: over oceans, deserts, mountains, forests, ice, and every ecosystem in between, and they have a profound effect on climate, and they are mostly naturally occurring.

The IPCC WG1, AR5 report (2013), Table 7.1, indicates that about 96% of aerosols are of natural origin, with the remaining 4% caused by humans. Adam Voiland, (2010, *NASA Earth Observatory*), put these figures at 90% and 10%. In any case, it is clear that the bulk of aerosols are from natural sources.

The two most abundant aerosols are sea salt and mineral dusts. They account for about 80% of the total aerosols in the atmosphere. Sea spray alone contributes about 50% of the total aerosol count; it is produced at the sea/ocean surface by bubbles bursting. Wind speed, sea-state, atmospheric stability, temperature and composition of the sea water contribute to the emission of sea salt into the atmosphere. Salt particles tend to reflect all the sunlight they encounter. Oceans cover more than 70% of the Earth's surface and above all that water is a huge amount of sea spray. When bubbles burst, chemicals escape into the atmosphere as *solar reflective* aerosols; i.e., they have a cooling effect on the climate.

But, in spite of the dominance of sea spray aerosols, the IPCC WG1, AR5 says,

"Our... process-based estimates of the total mass and size distribution of emitted sea spray particles continue to have large uncertainties."

It has *"large uncertainties"* about 50% of the aerosol effect on climate – not very encouraging, particularly considering the sensitivity of temperature to small albedo variations discussed in section 4.1.1.

Mineral dust, accounting for about 30% of total aerosols, is soil and/or sand particles that the wind blows into the atmosphere. As with sea salt, wind speed and strength are major determinants of the amount of particles that reach the atmosphere. According to NOAA, nearly all the planet's dust (about 94%) originates in the Northern Hemisphere, with about half of it originating in North Africa and, it says, North Atlantic Oscillation phases and El Niño events correspond to greater Saharan dust transport across the Atlantic. **So, we have naturally produced aerosols, transported around the world by naturally occurring wind and oceanic oscillations, affecting climate in a manner not fully understood by the IPCC et al!**

Human-made aerosols come from a variety of sources. Fossil fuel combustion produces large amounts of sulfur dioxide, which reacts with water vapor and other gases in the atmosphere to create sulfate aerosols. Automobiles, incinerators, smelters, and power plants are prolific producers of sulfates, nitrates, black carbon, and other particles. Deforestation, overgrazing, drought, and excessive irrigation can alter land surfaces, increasing the rate at which dust aerosols enter the atmosphere. Biomass burning, a common method of clearing land and consuming farm waste, produces smoke that consists mainly of organic carbon and black carbon. Though less abundant than natural forms, anthropogenic aerosols can dominate the air downwind of urban and industrial areas, i.e., they can have a significant *regional* effect.

As with clouds, different aerosols scatter or absorb sunlight to varying degrees, depending on their physical properties. Most aerosols are cooling, they reflect the Sun's energy by reflecting sunlight back out into space. In fact, there is only one aerosol, soot, also known as black carbon, that contributes to warming.

According to an article by Jeff Tollefson (15 January 2013, *Nature*), the contribution of soot to global warming is much higher than previously thought. A four-year study, published online by the *Journal of Geophysical Research*, roughly doubles most of the previous estimates of the warming that occurs when black carbon particles absorb solar radiation, heating up the atmosphere which melts snow and ice which, in turn increases heat absorption at the surface, a positive feedback effect. Note, white sheets of ice reflect a great deal of radiation, resulting in a negative feedback, cooling effect. According to the study, **black carbon's warming impact on the climate is roughly two-thirds that of carbon dioxide**. The study doesn't say so but these findings appear to challenge the IPCC et al dogma of climate sensitivity to CO_2 doubling – with a greater influence on the climate by aerosols (in this case, black carbon), CO_2 must have a smaller impact – it's a zero-sum balance sheet.

"Although many scientists had suspected that global climate models underestimated the role of black carbon, the magnitude of the impact has surprised many of the report's authors,"

says David Fahey (2013), an atmospheric scientist at NOAA in Boulder, Colorado, and a lead author in IPCC, AR4 WG1, 2007. (Dr. Fahey is Director of the Earth System Research Laboratory Chemical Science Division, NOAA).

Black carbon can alter reflectivity by (1) depositing a layer of dark residue on ice and other bright surfaces, and (2) by combining with falling snow flakes reducing the snow reflectivity. These effects are illustrated in Figure 4.5-6. The image on the left hand side depicts a situation with little or no soot, resulting in high reflectivity – a cooling effect. The image on the right depicts the opposite situation; a high soot concentration results in radiation absorption – a warming effect.

Figure 4.5-6 Effects of Soot on Ice (Source: NASA)
Left: Little to no Soot – high reflectivity
Right: High Soot Concentration – high absorption

In a 2003 article by Hansen and Nazarenko, they suggested that,

"The effect of soot on snow albedo (reflectivity), not included in climate studies such as those by the Intergovernmental Panel on Climate Change, is important.... A soot content of only a few parts per billion (ppb) is needed to reduce snow albedo by 1%. We estimate that soot reduces snow albedo about 3% in Northern Hemisphere land areas, 1.5% in the Arctic, and 0.6% in Greenland. Climate simulations show that this modest albedo effect would cause a global warming that is more than a quarter of the warming observed in the past century." (nasa.gov/home/hqnews/2003/dec/HQ_03420_black_soot.html)

Recall the discussion in section 4.1.1 on the high sensitivity of temperature to small changes in albedo; decreasing albedo by 3%, which Hanson and Nazarenko say occurs, had a 3°C warming effect.

The article notes that soot's increased absorption of solar energy is especially effective in warming the world's climate, saying,

*"This forcing is unusually effective, **causing twice as much global warming as a carbon-dioxide** forcing of the same magnitude."* Yet,

according to Hansen and Nazarenko, **the effects of soot are** *"not included in climate studies;"* why is that? Could it be that increased temperature due to aerosols would decrease the claimed impact of CO_2? Too cynical?

Although most aerosols remain suspended in the atmosphere for short periods, typically between one to ten days (according to Table 7.2 of IPCC AR5 report), during that period, they can travel vast distances. Particles moving with the atmosphere will travel thousands of miles in a week. Dust plumes from the Sahara frequently cross the Atlantic and reach the Caribbean. Winds sweep a mixture of Asian aerosols, particularly dust from the Gobi Desert and pollution from China, east over Japan and toward the central Pacific Ocean. Aerosols permeate the atmosphere.

Because of a lack of information concerning the temporal and spatial distribution of aerosols, i.e., how long the various particles survive and their journey within the atmosphere, they are key contributors to the uncertainties in current climate studies. To compound these uncertainties when aerosols clump together, they form complex hybrid particles.

".... estimating the direct climate impacts of aerosols remains an immature science. Of the 25 climate models considered by the Fourth Intergovernmental Panel on Climate Change (IPCC), only a handful considered the direct effects of aerosol types other than sulfates." (Earthobservatory.nasa.gov/features/aerosols).

The problem with this is that sulfates (salts) are reflective while soot is absorptive; the latter contributing to global warming in a significant manner according to Hansen and Nazarenko et al. So, on the warming side of the ledger, **aerosols are mostly ignored thereby allowing CO_2 to appear more effective in terms of global warming than it would otherwise be.**

On the other side of the ledger, the cooling effects of aerosols, research shows that the IPCC has in the past overstated the effects. According to Lewis and Crok (2014, *A Sensitive Matter: how the IPCC buried evidence showing good news about global warming*), who were both expert reviewers of AR5, report that in AR5, **the IPCC acknowledges that previous estimates of aerosol cooling impact were overstated**, i.e., based on more recent observations, aerosols are less effective in cooling. This is very important to the debate, since aerosol cooling has been used by the IPCC et al as a reason for the observed leveling of temperatures from about 1998 through 2014; i.e., the *pause*.

Human-caused aerosol cooling, the IPCC et al has said, offset the continuing warming caused by increasing CO_2. If that is now not the case or even if the contribution of aerosol to cooling is reduced, as AR5 says it is, it follows that the *offset* has no explanation, except perhaps natural climate events, such as oceanic oscillations, which are also poorly understood.

The most important information garnered from researching aerosols is that (1) they are mostly naturally occurring; (2) they permeate the atmosphere, but human-caused aerosols are concentrated around industrial areas; i.e., they are most abundant in the northern hemisphere, (3) the *warming* aerosol, soot, appears to contribute significantly to warming from which one can infer that CO_2 has a reduced impact, yet it is not included in IPCC models, and (4) the *cooling* aerosols, used by IPCC et al to explain the stable average global temperatures for about 15 years (while CO_2 continued to increase linearly), is overstated.

Section 4.5 Summary: Clouds and aerosols are two of the most important and least understood elements in climate studies. Both clouds and aerosols reflect and absorb energy; they both include microscopic particles far too small for IPCC models to accurately simulate and forecast events (discussed further in Chapter 5), and; their interactions and feedback processes are poorly understood.

IPCC reports used terms such as *"daunting"* when acknowledging its lack of understanding of these elements, and it admits that aerosol-clouds interactions in its models are *"primitive."*

The IPCC et al lack of understanding of these forcings is alarming considering the impact they have on climate. For instance, temperature sensitivity to cloud cover is such that according to President Obama's Science Czar, Dr. John Holdren, **a 1% increase in cloud cover would decrease the surface temperature by 0.8°C, the entire warming amount attributed to CO_2 since the start of the Industrial Age**.

In a complicated feedback process, clouds and climate are tightly integrated; when the climate changes, clouds change; when clouds change, the climate changes. Generally, high clouds have a net warming effect on temperature, and low clouds a net cooling effect. Clouds are an integral part of the Earth's water (hydrological) cycle, collecting water from the surface (evaporation), transporting and precipitating it at various regions around the globe, having a cooling effect.

As with clouds, aerosols, which are 90 to 95% natural, are poorly understood and impossible to model in the IPCC global circulation models which have grid resolutions billions of times the size of the particles they attempt to simulate (see Chapter 5, Models).

IPCC AR5 (2013) reported that the cooling effect of aerosols has been overstated in past reports – this is significant because aerosols have been used to compensate for poor model performance (not being able to replicate the steady temperatures from 1998 through 2014) by claiming aerosols offset the warming effect of CO_2. It follows that since the cooling effect of aerosols on temperature was overstated, the claimed warming effect of CO_2 must have been offset by some other natural event, perhaps the Pacific Decadal Oscillation which turned negative around that time.

4.6: Greenhouse Gases:

Earth is the only planet in the solar system with an atmosphere that can sustain life. The blanket of atmospheric gases not only contains the air we breathe but also modulates the energy balance as discussed in section 4-1. This section discusses greenhouse gases, climate sensitivity (the effects of doubling CO_2 levels), and the carbon cycle which, together with the water cycle, maintains the Earth's equilibrium.

Essentially, a greenhouse gas is one that absorbs heat (infrared radiation - which slows the rate at which the Earth can cool), and radiates it back to Earth. Combined, these processes warm the surface and atmosphere. This increase in temperature resulting from greenhouse gases (GHG) is called the *greenhouse effect.*

Carbon dioxide is a greenhouse gas that has, in simulated models, been shown to be directly correlated with temperature. Physicist Dr. Andrew A. Lacis, et al (2010, *Atmospheric CO_2: Principal Control Knob Governing Earth's Temperature, Science*) of NASA Goddard Institute for Space Studies, illustrated the role of CO_2 in driving temperature **if no other natural events or feedbacks are considered**. They ran a climate model with the then current CO_2 values (around 390 ppm) and reported that the model was an accurate representation of the prevailing climate at that time. They then removed all of the CO_2 and other *non-condensing* GHGs (methane, nitrous oxide, etc.) from the simulated atmosphere; water vapor, a condensing GHG, was retained. Without CO_2, their model demonstrated that within 10 years the temperature dropped more than 30°C. While the temperature was falling the amount of water vapor declined, which cooled the climate further, which together with formation of ice and snow and their albedo feedback effect, the Earth approached an ice-covered state in which there was very little water vapor present (similar to that of the Poles today). This exercise demonstrated that: (1) there is a relationship between temperature and water vapor – never in dispute; (2) water vapor is the most potent GHG – never in dispute and, the primary purpose of the exercise; (3) CO_2 and temperature are directly correlated.

That CO_2 is directly correlated with temperature is not the issue however, **the issue is whether or not the relationship is the dominant climate driver** *in the presence of other natural events and feedbacks,* and clearly it is not or, as CO_2 increases linearly, so too would temperature; Chapter 3 shows it does not. It is just one climate forcing event in a complex Earth system, discussed in section 4.7.

Greenhouse gases are necessary for life to exist; it is generally reported that without the greenhouse effect, the Earth's average temperature would be about 33°C (92°F) colder and the planet would be uninhabitable. For example, the Moon is approximately the same distance from the Sun as is the Earth; it has no greenhouse gases (or atmosphere) and its temperature varies from about 100°C in the daytime to lower than minus 100°C at night. Atmospheric and greenhouse gases moderate our climate.

It is important to understand - I didn't - that while the IPCC, politicians, media, et al discuss greenhouse gases and the size of their effect on the Earth's temperature as *dogma,* a quote from Gavin Schmidt, Director of NASA, provides a perplexing insight,

*"The size of the greenhouse effect is often estimated as being the difference between the actual global surface temperature and the temperature the planet would be without any atmospheric absorption, but with exactly the same planetary albedo, around 33°C. **This is more of a "thought experiment" than an observable state....**"* (2010, *Taking the Measure of the Greenhouse Effect*).

A thought experiment? The 33°C greenhouse effect is just an explanation of what might possibly be true. But of course, it must be an approximation since he says that albedo is constant (*exactly the same*) which it is not and, as we saw in Section 4.1 temperature is highly sensitive to albedo. Talk about creating uncertainty. We also saw that the Lacis et al experiment indicated a 30°C difference, a discrepancy of 3°C from the "*thought experiment*" value – and remember the total warming in the climate debate is about 0.8°C.

In the same article, Schmidt said that "*small changes in trace gases such as carbon dioxide* **might** *make a difference.*" Not very compelling from a Director of NASA.

On December 7th, 2009, the U.S. Environmental Protection Agency (EPA) declared that carbon dioxide and other greenhouse gases are harmful to people and the environment. Earlier in 2009, in a *Daily Signal* article called *Man's Contribution to Global Warming*, Nicolas Loris referred to it as "*planet-killing carbon dioxide.*"

The EPA was obviously pandering to alarmists and more importantly at the time to the Obama administration: **carbon dioxide (CO$_2$) is not a toxic gas;** it plays a vital role in plant and animal processes, including photosynthesis (CO$_2$ + sunlight + Water → Plant food (carbohydrates) + Oxygen). Obviously, we need the oxygen, and plants need the carbohydrates. But it serves *the cause* to portray CO$_2$ as a toxic gas or as a *planet killer;* who doesn't want to reduce or eliminate toxic gases, and who wants to *kill the planet*? Unbelievable nonsense!

The major constituents, by volume, of *dry air* are Nitrogen (78%), Oxygen (21%), and Argon (0.93%). This leaves about 0.07% of the atmosphere for all other gases which are mostly greenhouse gases (GHG). GHGs are therefore *trace gases.* CO$_2$ occupies about 5% of GHGs, approximately 0.05% of the atmosphere. Water vapor, or humidity, can occupy up to 95% of GHGs, and from almost zero percent of the atmosphere up to 4.0 or 5.0% on a very hot humid day.

This raises the question: in a humid climate, in which H$_2$O occupies up to 5% of the air, by volume, how is that the total atmospheric volume doesn't exceed 100%, when N$_2$ = 78%, O$_2$ =

21%, and Argon is around 0.93%? The answer is that as the volume of water vapor increases it displaces or evaporates an equal volume of some combination of N_2 and O_2. CO_2 remains relatively constant.

Of the 0.05% total atmospheric CO_2, about 5% is anthropogenic, i.e., **human-caused CO_2 is a trace of a trace gas occupying about 0.003% of the atmosphere**.

Water vapor is the most powerful GHG, it is the most abundant, **and it is 99.9% natural**. While CO_2 constitutes about 0.05% of the atmosphere, around 400 ppm by volume., water vapor varies from zero (theoretically, but zero humidity is impossible) to around 4 to 5%, or 40,000 to 50,000 ppm.

That water vapor is the most abundant greenhouse gas is dissembled (kindest word I can think of) by politicians and media. For example, see the chart below that ABC News presented a few years ago on its website:

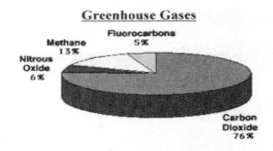

Greenhouse Gases

Fluorocarbons 5%
Methane 13%
Nitrous Oxide 6%
Carbon Dioxide 76%

This graph shows the distribution of GHG in Earth's atmosphere. Carbon Dioxide is clearly the majority.
www.abcnews.com/sections/us/global106.html

It states that *"Carbon Dioxide is clearly the majority"* greenhouse gas – it was a lie! It didn't say that the numbers were based on a (hypothetical) dry atmosphere, Yet, probably the public that viewed the website believed it to be true – why wouldn't they?

A more recent example of the deceit is in the 2017 EPA *"Overview of Greenhouse Gases,"* in which it presents the same pie chart (slightly different percentages), ignoring any reference to water vapor. (epa.gov/ghgemissions/overview-greenhouse-gases)

Table 4.6-1 was constructed from data published by the U.S. Department of Energy (DOE), summarizing concentrations of the various atmospheric greenhouse gases. While most likely accurate, it is misleading for the uninitiated observer, 99.438% of GHG may be CO_2 – yes, but again, no consideration for water vapor. Table 4.6-2 shows the composition of GHGs when water vapor is included; Methane, Nitrous Oxide, and Misc. (CFCs, etc.) are negligible.

Table 4.6-1
Greenhouse Gases (except water vapor)
U.S. Department of Energy Data, (October, 2000)

	Percent of Total
Carbon Dioxide (CO_2)	99.438%
Methane (CH_4)	0.471%
Nitrous Oxide (N_2O)	0.084%
Misc. gases (CFC's, etc.)	0.007%

Table 4.6-2
Atmospheric Greenhouse Gases
(man-made and natural) as a % of Total GHGs

	Percent of Total
Water vapor	95.00%
Carbon Dioxide (CO_2)	4.97%

From Table 4.6-2, we see that about **0.25% of the total greenhouse effect** (5% of 4.97) **is caused by humans** (water vapor is 99.9%

natural). The Kyoto Accord targeted a reduction in CO_2 emissions of 5% over the period 2008 to 2012, against 1990 levels, and by at least 18% below 1990 levels over the period from 2013 to 2020. So, a world intergovernmental panel (IPCC), media hysteria, politician scaremongering, hundreds of billions of dollars with commitments for trillions, and an attempt to impose draconian energy policies on the world – all to reduce anthropogenic CO_2 from (approximately) 0.25 to 0.21% (0.04%) by 2020!

A 2008 report by NASA says recent satellite data has validated the role of the water vapor as a critical component of climate change. The relative importance of water vapor to the total greenhouse effect however is difficult to determine: water vapor, i.e., humidity, is highly variable from day-to-day and from place to place, whereas CO_2 is fairly evenly mixed in the atmosphere, around 0.05%. This can be understood by comparing the greenhouse effect in Miami (high humidity; at sea level) to say Denver (low humidity; at 5,280 feet altitude): while both cities may have the same temperature during a day, on a clear night the temperature in Denver will drop substantially more than in Miami because the greenhouse effect of the water vapor in Miami retains more heat. **Water vapor is a more efficient greenhouse gas than CO_2.**

Water vapor amounts depend on local environments, the percentage within the atmosphere hinges on geology and temperature. Above the ocean in warm environments, the percentage of water vapor within the atmosphere can be as high as 4 or 5%; e.g., in the tropics. This is due to two effects. First, higher temperatures lead to more evaporation of local water into the atmosphere. Second, the amount of water that can be held within the atmosphere is higher at greater temperatures. In contrast, cold locations such as the geographic poles, and arid regions, such as deserts, have low water vapor levels, which is why deserts are very cold at night.

The effect of water vapor on the climate is simply ignored by most advocates of human-caused warming but not by the IPCC which in a 2007 Report wrote,

"Water vapour is the most abundant and important greenhouse gas in the atmosphere. However, human activities have only a small direct influence on the amount of atmospheric water vapour."

Interesting – has anyone heard politicians or media discussing water vapor in the context of climate change or global warming? Rhetorical, of course not.

In its models, the IPCC assumes that a small, manmade warming signal caused by CO_2 warms the atmosphere which leads to an increased water vapor concentration and, since water vapor is an efficient greenhouse gas, it amplifies the CO_2 forcing effect by a *factor of three*. **Without this level of amplification, there is no CO_2 global warming case.**

While most scientists accept amplification of the CO_2 signal occurs, many consider the IPCC multiplier to be too high by a factor of 2 or 3; i.e., it should be half or a third of the effect. And it's not just skeptics who consider the multiplier to be excessive: a NASA-funded study (2004, *Meteorological Society's Journal of Climate*) conducted by physicist Ken Minschwaner and Andrew Dessler a researcher with the University of Maryland and NASA's Goddard Space Flight Center, found that increases in water vapor as the Earth warms were lower than many climate forecasting models have assumed. Dr. Minschwaner, Professor of Atmospheric Physics at New Mexico Institute of Mining and Technology, said,

"Our study confirms the existence of a positive water vapor feedback in the atmosphere, but it may be weaker than we expected."

He didn't say how much weaker the multiplier is, but this is very important; the amplitude of the multiplier is the cornerstone of the AGW CO_2 claim.

Water vapor does, in fact, increase its volume with increasing temperature, and water vapor is an effective GHG, so if they were the only two factors involved it may seem to be a reasonable relationship, but climate is much more complicated. Without some sort of governor (negative feedback) this would result in an unrealistic scenario of CO_2 increasing temperature, which increases water vapor, which increases temperature, which increases water vapor, and so on, creating a runaway situation which of course is not what happens. The control, presented in many research papers, is that water vapor also produces clouds which, in addition to absorbing heat from the Earth's surface, a greenhouse effect, reflect incoming solar radiation back into space (albedo), a cooling effect, and precipitates out as rain (cooling).

Climate Sensitivity to CO_2: The IPCC (2007, *Fourth Assessment Report* (AR4)) defines climate sensitivity as *"the equilibrium global mean surface temperature change following a doubling of atmospheric CO_2 concentration."* The *equilibrium global mean surface temperature* is the average temperature after a time sufficiently long for both the atmosphere and the oceans to respond to the stimuli, in this case a doubling of CO_2. While land equilibrium is reached fairly quickly, it can take decades or centuries for the oceans to do so. The AR4 continued,

"Climate sensitivity is largely determined by internal feedback processes that amplify or dampen the influence of radiative forcing on climate."

What it doesn't say is that the feedback used in its models is mostly *positive* feedback, i.e., amplifies (exacerbates) the effect of CO_2. Without assuming CO_2 amplification, it is generally agreed by both sides of the debate that the sensitivity, i.e., doubling CO_2

concentration, on global mean temperature would be around 1°C, which is not alarming. By including its *amplification* factor (x3) however, IPCC AR4 assumed a sensitivity of 2°C - 4.5°C. IPCC AR5 (2013) reduced the lower limit back to 1.5°C where it was since the first report in 1990. But remember **the big warming number (4.5°C) come not from measurements but from computer models which** as discussed in section 4.5 and in Chapter 5 **are** *fraught* **with uncertainties** and riddled with assumptions and mathematical *techniques* that allow modelers to obtain any result they wish.

Possible *negative* feedback scenarios such as one postulated by Lindzen et al (2001, *Does the Earth have an Adaptive Infrared Iris, Bulletin of the American Meteorological Society*), which would dampen warming (below the 1°C) rather than enhance it, have not been adequately considered by the IPCC. Dr. Lindzen is one of the most distinguished scientists in the world: a past professor at the Universities of Chicago and Harvard; the Alfred P. Sloan professor of meteorology at MIT; a member of the National Academy of Sciences; and a lead author of the 2001 IPCC report (prior to his resigning as discussed in Chapter 2). Dr. Lindzen readily acknowledges, as do all the scientists whose work I have reviewed, that temperatures have raised in the 20th century and that increased CO_2 can raise temperature. But,

"none of those well-established facts" he argues, *"can justify the doomsday case, or the doomsday solution, and certainly cannot justify twisting science for political goals."*

According to Lindzen and Choi (2009, *On determination of climate feedbacks from ERBE data, Geophysical Research Letters*), a doubling of atmospheric CO_2 and taking all feedback into account, positive and negative, would result in a rather modest warming of only 0.5°C – 0.7°C. This, they attribute in part to what they call the *Iris Effect*. The Iris Effect hypothesis is basically that a warmer climate increases evaporation and precipitation (cooling effect), reducing the number of water molecules available to form cirrus clouds, so the number of clouds decrease and, with less high cloud cover more infrared radiation can escape to space (additional cooling), thereby creating a strong climate-stabilizing, negative cloud feedback.

In a follow-up research project by Lindzen and Choi (2011, *Asia-Pacific J. Atmos. Sci.*), in response to peer critiques of their 2009 article, they confirmed their hypothesis of the Iris Effect and the understated effect of negative feedback in IPCC models.

As discussed in section 4.1, much of the near-infrared, visible and ultraviolet light from the Sun passes straight through the atmosphere and warms the Earth's surface (land and oceans). The warm surface emits infrared light (at varying wave lengths depending on surface temperature conditions) that gets absorbed primarily by water vapor and carbon dioxide, eventually being re-emitted (in all directions). Much of this emission is downwards which, in addition to slowing the emission of heat into space makes the surface of the Earth warmer than if it had no greenhouse effect. So far there is no dispute. And it would seem to follow that the more greenhouse gases (GHG) there are in the atmosphere the more radiation they absorb and re-radiate back to Earth with a concomitant increase in temperature. This is the fundamental thesis of AGW advocates.

An important point about CO_2 however, is that adding more of it to the atmosphere beyond a certain amount produces only a small effect on warming. There are two reasons: (1) CO_2, as well as other GHGs, can only absorb infrared energy at certain wavelengths and there are only so many wavelengths with absorbing capacity available for each gas, and; (2) adding more CO_2 to the atmosphere does not have a linear effect on absorption, rather it is more logarithmic; it has a limit to which it approaches in ever decreasing amounts; i.e., asymptotically.

(1) Figure 4.6-1 illustrates the absorption wavelengths for CO_2 and water vapor (H_2O). The figure has a horizontal scale from 1μm to 30 μm (micrometers-millionths of a meter) representing wavelengths, with the vertical scale from 0% to 100% where zero means no absorption and 100% means complete absorption; i.e., that band is saturated and can hold no more of the particular gas; no more absorption can occur, no matter how much more of the gas might be added to the atmosphere. We can see from the figure that three of the available bands for CO_2 are saturated leaving only two possible bands, at about 2μm and 5μm, the former band being totally overlapped by a H_2O band so is less effective (water vapor, H_2O, is a more efficient GHG than CO_2). Although it appears that water vapor is also saturated in nearly all its bands, the figure cuts off at 1μm but there are several more bands at lower wavelengths available to absorb water vapor which is not the case for CO_2.

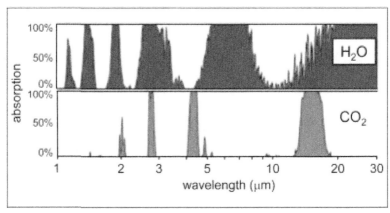

Figure 4.6-1 Absorption bands for H2O and CO_2 gases. (Source: Public domain)

(2) The pre-industrial CO_2 level was around 280 ppm; today it is approximately 400 ppm, a 42% increase. Over that period, the average global temperature increased about 0.8°C. If there was a linear relationship between CO_2 and temperature the additional 58%, to double the pre-industrial CO_2 amount to 560 ppm, would

increase the projected mean temperature to about 1.9°C. Incidentally, according to NOAA data presented in Chapter 3, Figure 3-11, when extrapolated at the same rate shows that it will take more than 100 years to reach 560 ppm.

But since the absorption of CO_2 is effectively logarithmic, the 120-ppm increase, from 280 to 400 ppm had a far bigger impact on temperature than the remaining 140 ppm, from 400 to 560 ppm, will have. **The majority of the temperature increase due to CO_2 is behind us;** approximately 70% of the warming due to doubling the pre-industrial level of CO_2 has already happened. An analogy of a logarithmic effect is as follows:

Consider that you have several pairs of sunglasses, each of which blocks 50% of the light. If you put two pairs on, the second pair blocks 50% of the 50% that the first pair allowed to pass through, i.e., 25% of the incoming light is passed through. A third pair would block 50% of the 25% allowing 12.5% to pass, and so on, with each additional pair having less effect until eventually there is no appreciable effect. This relationship between sunglasses and light absorption is effectively logarithmic. Similarly, the effect of adding more GHG to the atmosphere with its concomitant increase in temperature, is not linear; just as adding the second pair of sunglasses does not absorb as much light as adding the first pair does. In practical terms, this means that adding the first portion of a GHG produced a large effect, but further additions have less and less effect.

Idso et al (2016, *Why Scientists Disagree about Global Warming*) says that **the IPCC ignores mounting evidence that climate sensitivity to CO_2 is much lower than its models assume,** suggesting it should be around 1°C, and lists 27 peer-reviewed research articles that support this claim. This is consistent with the NASA-funded study showing that water vapor is less of an amplifier than that used in climate models. Of course, these findings do not support *the cause* and so are not reported in mainstream media.

Carbon Cycle: Carbon is essential to life as we know it on Earth; it is connected to everything that matters to us, our bodies, ecosystems, climate, and the health of the planet. According to Dr. David Archer (2010, *The Global Carbon Cycle*),

"Over millions of years, the carbon cycle acts to stabilize and regulate climate. Changes in the geometry of the Earth's surface, driven by plate tectonics, push the "set-point" of the CO_2 thermostat up and down, allowing Earth's climate to drift slowly from hothouse to icy states. But the carbon cycle is functioning as a regulator of climate, a stabilizing negative feedback."

The amount of carbon dioxide in the atmosphere is maintained through a balance between processes such as photosynthesis, respiration and natural combustion. Green plants remove carbon dioxide from the atmosphere by photosynthesis – oxygen is a by-product. Living organisms, including all plants and animals, release energy from their food by respiration - CO_2 is a by-product. Respiration and combustion both release CO_2 into the atmosphere. These however, are a small part of the overall carbon cycle.

Carbon continuously cycles through the major components of the Earth system driven by processes that occur at vastly different time scales, from fractions of a second (photosynthesis) to millions of years (formation of fossil fuels). There are four major components of the Earth system, called *spheres*. They are the biosphere, atmosphere, hydrosphere, and the geosphere (also known as the lithosphere). Each contains reservoirs of carbon (sinks) that are exchanged between and among the various spheres.

The biosphere is the boundary between the atmosphere, hydrosphere, and geosphere. It is the collection of all living things on Earth: plants, animals, bacteria, fungi, algae, and single-cell organisms, and includes the relatively thin layer of soil that covers parts of the Earth's surface. **There is more carbon in soil than in the total amount in the atmosphere and the rest of the biosphere combined**. Within the biosphere, are areas such as deserts, grasslands and tropical rain forests.

The atmosphere is the blanket of gas, air that surrounds the Earth, comprising nitrogen, oxygen, argon, the two most significant greenhouse gases, water vapor and carbon dioxide, plus numerous other trace gases. The level of atmospheric CO_2 in the atmosphere depends on constantly changing equilibrium between its sources and sinks and, according to Jaworowski, et al (1992, *Atmospheric CO_2 and global warming: a critical review*), flows of CO_2 between the oceans and the atmosphere are so important to the CO_2 budget that even **very small natural fluctuations of CO_2 in this process can mask anthropogenic CO_2 inputs into the atmosphere**.

The hydrosphere is all of Earth's bodies of water, including oceans, lakes, rivers, and even the groundwater and frozen water (snow, frozen ponds, glaciers, sea ice, etc.); it is the second largest reservoir. According to NASA (*The Ocean's Carbon Balance*), Archer (2012), et al, **one of the largest unknowns in understanding the greenhouse effects is the role of oceans as a carbon sink**. Professor Plimer of the School of Earth and Environmental Sciences at the University of Adelaide (2009, *Heaven and Earth*), says that water and evaporation are the *regulators* of global climate and that an increase in heat from ocean evaporation for a very small temperature rise (0.3°C) is more than enough to offset a doubling of atmospheric CO_2. According to Plimer et al, **oceans**, which cover more than 70% of the Earth's surface, **have the capacity to absorb all the fossil fuel that exists in the geosphere**; i.e., the total amount of fossil-fuel deposits in the Earth.

The geosphere includes the solid part of Earth, mountains, valleys, soil, bedrock, etc., but the major geosphere reservoir is in limestone in the Earth's crust. It is by far the largest reservoir of carbon and the slowest to affect the carbon balance. The surface of the geosphere, where the rocky part of the planet is in contact with water, air, soil, and vegetation is generally where the spheres intersect and affect each other.

Matter and energy are always circulating among the spheres, connecting them in many ways. Sometimes this circulation can happen very quickly such as when a volcano erupts, moving tons of tiny particles, including CO_2, from the geosphere into the atmosphere. Other examples of circulation happen much more slowly, such as when a fallen tree decays and carbon move from the biosphere to the geosphere (becomes fossil fuel).

Total human CO_2 emissions, primarily from use of coal, oil, natural gas, and the production of cement, are currently about 8.6 giga tons (Gt) per year (giga - billions). To put these figures in perspective, numerous sources estimate that the atmosphere contains about 800 Gt; the world's oceans (hydrosphere) contain 39,000 Gt; vegetation, soils, and waste or debris of any kind (biosphere) contain around 2,000 Gt; and rocks such as limestone in the Earth's geosphere estimates range from 65,000,000 Gt to 100,000,000 Gt of carbon. **Each year, the ocean surface and atmosphere exchange** an estimated 90 Gt, **ten times more carbon than humans produce** (90 GT vs. 8.6 GT); vegetation and the atmosphere exchange about 110 Gt.

The movement of geosphere tectonic plates causes CO_2 to be continuously released, from volcanoes and earthquakes. When volcanoes erupt, they vent gas, including CO_2, into the atmosphere and, following earthquake eruptions many fault zones release water vapor and CO_2 into the atmosphere from these fractures. Interestingly, according to Plimer (2009), even though there are more than 10,000 earthquakes each year, out-gassing CO_2 into the atmosphere, this source of natural CO_2 is not considered in IPCC models: interesting, but not surprising; the more *natural* CO_2 there is in the atmosphere, the less that can be attributed to humans. The carbon cycle has fast and slow components.

The Fast Carbon Cycle: The fast carbon cycle is largely the movement of carbon through life forms on Earth, or the biosphere, interacting with the atmosphere.

Plants and phytoplankton (microscopic organisms in the ocean) are the main components of the fast carbon cycle which take carbon dioxide from the atmosphere by absorbing it into their cells. Using energy from the Sun, both plants and plankton combine CO_2 and water to form carbohydrates and oxygen - *photosynthesis*. The carbon is returned to the air through *respiration* by plants and animals that have eaten the plants, decay, and decompose after the plants die.

The fast carbon cycle has a strong annual cycle and varies with the seasons: carbon dioxide concentrations in the air decrease during spring and summer when plants are growing and increase during winter when plants die or go dormant. **The stabilizing effects (negative feedback) of photosynthesis to climate have been underestimated by IPCC models** as revealed by the following research reports.

A report from the Daily Telegraph, June 2017, By Sarah Knapton, Science Editor says,

*"**Global warming may not be damaging the Earth as quickly as feared** after scientists found that **plants can soak up more carbon dioxide than previously thought**. According to researchers, climate models have failed to take into account that when carbon dioxide increases in the atmosphere, plants thrive, become larger, and are able to absorb more CO_2."*

Any system that has survived for more than 4 billion years, such as the Earth, is an adaptable one. From Dr. Ying Sun, Assistant Professor of Geospatial Sciences at Wyoming University (2014, *Proceedings of the National Academy of Sciences*),

'The terrestrial biosphere may absorb more CO_2 than previously thought.... The team estimates that climate scientists have underestimated the ability of plants to grow and absorb carbon dioxide by as much as 16 per cent.... **It was originally thought that vegetation on Earth currently removes one quarter of all human emissions. But the new study suggests it is far higher."**

This is but one example of the Earth's adaptability to its environment; i.e., stabilizing negative feedback, discussed further in section 4.7. And from Zhu et al (*2016, Greening of the Earth and its drivers, Nature Climate Change*),

"From a quarter to half of Earth's vegetated lands has shown significant greening over the last 35 years largely due to rising levels of atmospheric carbon."

The above research results are further evidence that the Earth is a self-correcting system. According to David Archer professor of geophysical sciences at the University of Chicago, et al, while humans are adding fossil fuel carbon, the natural carbon cycle is absorbing about 50% of the CO_2 emissions into the oceans and land surface (soils). It is also being absorbed by plants. **The carbon cycle is moderating the climate impacts of anthropogenic CO_2 release**, acting as a stabilizing negative feedback, particularly the vast absorption capacity of the oceans.

The Slow Carbon Cycle: Over the long term, the carbon cycle maintains a balance that helps keep Earth's temperature relatively stable, similar to a thermostat. This *thermostat* works over a few hundred thousand years, as part of the slow carbon cycle. For shorter time periods, tens to a hundred thousand years, the temperature can and does vary naturally. And, as discussed in Chapter 3, Earth swings between glacial and interglacial periods on these time scales.

An interesting theory that highlights the interactivity of the Earth's continuing evolution is that the uplift of the Himalaya Mountain range, due to tectonic plate collision (See section 4-2), beginning about 50 million years ago, reset the Earth's *thermostat* by exposing a large source of fresh rock to the atmosphere to absorb more carbon into the slow carbon cycle. You may recall from section 4-2 that this also had a significant effect on wind patterns and rainfall in the region and, consequentially, climate.

The movement of carbon from the atmosphere to the geosphere (rocks) begins with rain. Atmospheric carbon combines with water to form a weak acid, carbonic acid, which falls to the surface in rain. The acidic rain dissolves rocks (into soil), a process called *chemical weathering*, and releases various chemicals into the ocean where the carbon is eventually stored in limestone. According to Plimer, weathering regulates the Earth such that there is no permanent icehouse and no runaway greenhouse, and he says, "*the carbon cycle is essentially driven by solar energy via the water cycle.*"

Most of Earth's carbon is stored in rocks in the land and ocean crusts and mantle (lithosphere); these de-gas (emit CO_2 into the atmosphere) over geological time periods from mid-ocean ridges, volcanoes, earthquakes, and tectonic subduction where plate crusts collide (see section 4-2, Tectonics, for detail). When certain plates collide, one sinks beneath the other (subduction) and the rock it carries melts under the extreme heat and pressure, releasing carbon dioxide.

The slow carbon cycle also contains a faster component, the ocean-atmosphere interface. At the surface of the oceans, where air meets water, CO_2 dissolves in and ventilates out in a steady exchange with the atmosphere. Winds (atmospheric circulation), currents (oceanic circulation), and temperature control the rate at which the oceans and carbon dioxide interact with the atmosphere.

The Global Conveyer Belt, described in section 4.3, plays a key role in the carbon cycle; cold water at the poles absorbs CO_2 which, about 1,000 years later, is released into the atmosphere (along with water vapor) mainly at the tropics where cold water upwells. Note, unlike many substances, such as salt, sugar, etc., CO_2 is more absorbent in cold water than in warm water so that in the polar areas, the cold water absorbs more CO_2 than elsewhere.

The above descriptions notwithstanding, according to Hoffman and Simmons (2008, *The Resilient Earth*), et al, for more than three decades scientists studying the global carbon cycle have identified an imbalance in the carbon budget, called the *missing sink*. Measurements show that only about three-quarters of the CO_2 being produced (natural and human-caused) was accumulating in the atmosphere and oceans. The remainder, about 25%, was presumably captured on land, **but no one *knows* where it is going**. Nine years later, Li Yu et al (2017, *Nature, Geoscience*) say *"there is still a lot of uncertainty about how carbon sinks work."*

Many hypotheses have been presented including an adaptive biosphere as described above where plants adjust and absorb greater amounts of CO_2. Hoffman and Simmons say that scientists' *best guess* is that carbon is being absorbed by undisturbed or resurgent ecosystems, but they don't know exactly where they are. Sjogersten et al (2014, *Tropical wetlands: A missing link in the global carbon cycle?*), say that **tropical wetlands are not included in Earth system models, despite contributing large amounts of CO_2 emissions** from land use, land use change, and forestry in the tropics. Their research identified a *"remarkable lack of data on the carbon balance and gas fluxes from undisturbed tropical wetlands."* This is important because they say that this lack of knowledge about the carbon cycle *"limits the ability of global change models to make accurate predictions about future climate"* which is *Newspeak* for "the models are inaccurate; they don't work."

A key component of IPCC scaremongering stems from its assertion that anthropogenic CO_2 remains in the atmosphere from 50 to 200 years – note the typical IPCC margin of uncertainty! According to Plimer (2009), numerous studies show CO_2 lifetime in the atmosphere to be around 5 years, a number, Professor Plimer says, was acknowledged by a former IPCC chairman, Bert Bolin.

Professor Plimer says that the considerable difference in atmospheric CO_2 lifetime, supported by 37 independent measurements and calculations using six different methods compared to that used in IPCC models, resulting in the difference between 5 to 10 years versus 50 to 200 years, has not been explained by the IPCC. He says that the assumptions made by the IPCC in its models are incorrect (the 37 studies are listed in Lawrence Solomon's 2008 book, *The Deniers*, pp 82-83).

Professor Tom Segalstad, head of the Mineralogical-Geological Museum at the University of Oslo (1997, *Carbon Cycle Modeling and the residence Time of Natural and Anthropogenic Atmosphere CO_2*), says that until the *IPCC-influenced science era*, **the world of science was near unanimous that CO_2 couldn't stay in the atmosphere for more than about five to ten years** because of the oceans near limitless ability to absorb it. He says that those who claim that CO_2 lasts decades or centuries have no measurements or physical evidence to support their claims; nor have they shown the 5 to 10-year view to be wrong. Although the onus of proof of a hypothesis is on the proposer of the hypothesis, the 5 to 10-year comment refers to the IPCC's lack of consideration of alternative scientific findings.

This is potentially massive: Dr. Plimer says that **if CO_2 lifetime is 5 years, then the amount of total atmospheric CO_2**

derived from burning fossil fuel would be about 6% of that assumed by the IPCC, a 94% reduction! These are startling statistics that should at a minimum create some degree of uncertainty among alarmists. And while the IPCC retains a 95% confidence in its model projections, there is a little less hubris by other government agencies expressing their uncertainties.

According to the U.S. Department of Energy (DOE), *Genomic Science Program* (2010),

"The global carbon cycle plays a central role in regulating atmospheric carbon dioxide levels and Earth's climate, but **knowledge** *of the biological processes operating at the most foundational level* **of the carbon cycle remains limited.... Even minor changes in the rate and magnitude of biological carbon cycling can have immense impacts on whether ecosystems will capture, store, or release carbon.** *Developing a more sophisticated and quantitative understanding of molecular-scale processes that drive the carbon cycle represents a major challenge, but* **this understanding is critical for predicting impacts of global climate change.**" (genomicscience.energy.gov/carboncycle).

Earlier we saw the results of more recent research radically increasing biosphere absorption of CO_2 by 16%.

There is about a 25% discrepancy between the Earth system model and observed atmospheric CO_2 that cannot be accounted for (the missing sink). The DOE expresses uncertainty in the biological part of the carbon cycle; NASA expresses uncertainty in its understanding of the role of oceans in the carbon cycle; the IPCC ignores earthquakes and sub-surface volcanoes that continuously emit CO_2 into the atmosphere and, except for its convenience as an amplifier, the IPCC generally ignores the most effective greenhouse gas, water vapor, and finally; without the IPCC's unverified 50 to 200 year residence time for CO_2 in the atmosphere, the natural carbon cycle easily accommodates the minor perturbation (forcing) due to anthropogenic CO_2.

Section 4.6 Summary: Carbon is essential to all life on Earth; it is connected to everything that matters to us, our bodies, ecosystems, climate, and the health of the planet. Carbon dioxide is an essential gas for all animals and plants.

CO$_2$ is a trace of a trace gas, and human contribution, at about 0.003% of the atmosphere volume, is miniscule in comparison to the volume of the atmosphere and the annual exchanges of CO_2 within the carbon cycle.

Most media ignore water vapor when classifying greenhouse gases even though it comprises up to 95% of greenhouse gases. Without a water vapor *amplifier* that increases the (positive feedback) effect of CO_2 threefold however, carbon dioxide could not possibly raise temperatures to the levels predicted by IPCC climate models. And IPCC models minimize negative (stabilizing) feedback that is obviously occurring; otherwise, there would be a runaway situation, which there is not.

The relationship between CO_2 and its heat absorption capacity is logarithmic, that is, the initial addition had the greatest effect with additional amounts having a diminishing effect, until eventually adding more CO_2 will have virtually no effect. The Earth has already experienced about 70% of the effects of doubling the amount of CO_2 from pre-industrial levels.

Atmospheric CO_2 is a tiny fraction of the carbon available within the Earth's systems and carbon cycle whereby carbon is circulated among the biosphere, atmosphere, hydrosphere, and geosphere. Over geologic time, the carbon cycle established the Earth's thermostat (e.g., Himalayan Mountains), and then in real-time, along with other natural processes including the water cycle, photosynthesis, etc. maintains a balanced environment.

Scientists agree that they don't fully understand the carbon cycle and specifically they cannot explain a 25% variance in the amount of CO_2 that they consider should be in the atmosphere against the amount they measure; this is called the *missing sink*. Without knowing where this CO_2 is absorbed, models cannot be predictive.

As in other fields of climate science (clouds, aerosols, etc.) there are uncertainties and lack of knowledge about many processes associated with the carbon cycle, both the fast and slow cycles. Specifically, IPCC models underestimate the corrective absorption (negative feedback) of plant life in the fast carbon cycle which are adapting to increased CO_2 levels. And there is an accepted lack of knowledge by U.S government agencies: DOE about the biological process within the fast carbon cycle and NASA's uncertainty about the atmosphere/ocean carbon exchange process in the slow cycle.

4.7: The Earth System:

Everything that happens on the planet is part of a complex set of interactions within and between air, water, land, and life. The stability of a system is intimately connected to its equilibrium state. If a system in equilibrium is disturbed then, if it is stable it tends to return to or oscillate about its original equilibrium state. An unstable system tends to continue to move away from its original equilibrium state when disrupted from it. The Earth is a stable system and, in accordance with Le Chatelier's principle, which says that feedback mechanisms in stable systems are negative, it is adaptable and self-correcting. This section presents a high-level view of the Earth system and subsystems, followed by a discussion of feedbacks within and between subsystems.

Figure 4.7-1 is a macro depiction of the Earth system. It shows the relationships between climate forcings and the Earth's subsystems or *spheres*, cycles of energy and matter, and events that temporarily disrupt equilibrium.

Sections 4.1 through 4.6 discussed individual climate forcings, external and internal: The most dominant external forcing or source of energy (basically the Earth's engine) is the Sun; the other external energy source is cosmic rays emitted from outside the galaxy. While being modulated by the atmosphere and greenhouse gases, the Earth's temperature is driven by the intensity of the Sun's radiation at the top of the atmosphere, a function of its solar activity, the Earth's orbit (distance from the Sun), and planetary constellations. Its tilt determines the distribution of energy resulting in winds and currents, whose circulation patterns are a function of pressure and temperature variation and the Earth's rotation. Cosmic rays penetrate the atmosphere and create clouds; and clouds affect climate.

Internal climate forcings, shown in Figure 4.7-1 as wind and ocean circulation, oceanic oscillations, clouds and aerosols, and greenhouse gases are affected by the energy from the Sun interacting with the Earth's spheres and climate feedback mechanisms. Tectonics, the continuous movement of land and oceans is energized from below by radioactive decay of heavy metals in the Earth's core (such as uranium) which heats the mantle, upwelling magma which moves tectonic plates creating, in the short-term volcanoes, earthquakes, and tsunamis, and over geologic time changes mountain structures, continent boundaries, and ocean seabed topology.

Earth System

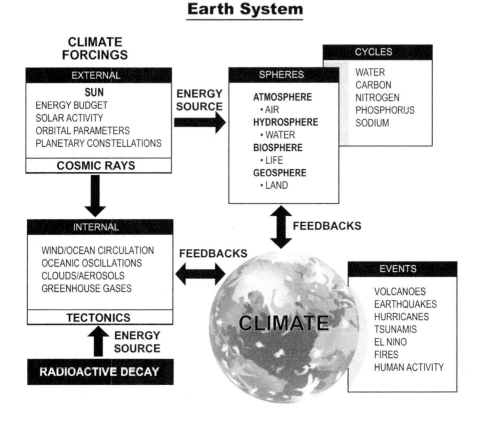

Figure 4.7-1 Earth System – A Macro View

Earth's System includes the interactions between the physical, chemical, and biological components of four major spheres: the atmosphere (air – nitrogen, oxygen, argon, and greenhouse gases; protects life from harmful radiation, warms the planet to habitable levels, transports thermal energy around the globe; hydrosphere (water – oceans, lakes, rivers, glaciers, and snow/ice; evaporation, condensation and precipitation maintain the Earth's energy balance); biosphere (life – forests, grasslands, animals, and vegetation; photosynthesis), and; geosphere (core, mantle, crust, and surface rocks and soil).

These spheres are interconnected by *cycles*, which, over time, store, transform and/or transfer matter and energy throughout the Earth system in ways that are governed by the laws of conservation of matter and energy; i.e., matter and energy cannot be created or destroyed, they can only be changed or cycled through various forms. For example, during the evaporation of water (water cycle), matter changes from liquid to gas, and during photosynthesis (carbon cycle) light energy converts to chemical energy. While there are numerous cycles essential for life to exist (nitrogen, phosphorus, sodium, etc.), the water and carbon cycles are the most influential on climate.

Events are internal phenomenon that can occur naturally, such as earthquakes or volcanoes, or can be caused by humans, such as an oil spill or deforestation, etc. An event can be the *cause* of changes to occur in one or more of the spheres, or an event can be the *effect* of changes in one or more spheres. This two-way cause and effect relationship between an event and a sphere is called an *interaction*. Interactions occur between and among spheres; for example, a change in the atmosphere can cause a change in the hydrosphere, and vice versa; a change in the hydrosphere (H) can cause changes in the atmosphere (A), biosphere (B), and/or the geosphere (G). While events disturb the steady-state conditions of the climate, they are also an intrinsic part of the climate.

Interactions often occur as a series of chain reactions whereby one interaction leads to another, which leads to another, and so on, resulting in a ripple effect through the Earth's spheres. For example, a volcanic eruption (an event in the geosphere) releases a large amount of particulate matter into the atmosphere (A). These particles (aerosols) serve as nuclei for the formation of water droplets which combine in the atmosphere to form clouds (H). Wind (A) moves clouds around the globe until they cool and precipitate. Rainfall (H) often increases following an eruption, stimulating plant growth (B). Particulate matter in the air (A) falls, initially smothering plants (B), but ultimately enriching the soil (G) further stimulating plant growth (B) which eventually (millions of years later) form fossil fuels (G). The new vegetation

(B) affects the Earth's energy balance by changing its albedo and increases photosynthesis which affects the amount of atmospheric CO_2 (A), and so on.

A lightning strike (event) may ignite a forest fire destroying all the plants in an area (B), polluting the air (A) with aerosols and gases, which affect clouds (H) and atmospheric albedo/absorption properties (A). The absence of plants leads to an increase in erosion (washing away) of soil (G) and an increase in CO_2 (decrease in photosynthesis). Increased amounts of soil entering streams increased turbidity, or muddiness, of the water (H). Increased turbidity of stream water can have negative impacts on the plants and animals that live there (B), and changes in the water albedo (H) which impacts the energy budget. Heat from the fire removes moisture from the air (A), soil (G), and vegetation (B) through the process of evaporation. The intense heat from the fire may cause some rocks to break apart (G), and so on.

Earth System Feedbacks: As discussed in Section 4.1, for a relatively steady temperature the amount of incoming solar energy absorbed by the climate system must be balanced by an equal amount of outgoing radiation at the top of the atmosphere. This is accomplished by the interactions and cycles within and between the Earth's spheres and, perhaps more importantly, the feedback loops that occur within and between them.

Sphere interactions and cycles include numerous feedback loops, positive and negative, with the net effect being negative, i.e., stabilizing. If this were not the case, the energy balance would rise or plunge, in a runaway sequence depending on the direction of the *positive* forcing event; we would either boil or freeze.

Feedback loops include short-term and long-term. An example of a short-term *negative* or stabilizing feedback loop is within the water cycle; the Sun heats the Earth surface, water evaporates, rises, cools, and condenses, forms clouds that precipitate and cool the surface. An example of a short-term *positive* or destabilizing feedback loop is when the Sun heats ice, the ice melts exposing lower albedo surface, land or ocean, which absorbs heat, which melts ice, exposing more surface, etc.

Those are simple feedback mechanisms; a more intricate feedback loop involves water vapor and lapse-rate (the rate at which atmospheric temperature decreases with increasing altitude). One theory is that since the air cools with increasing altitude, it emits less and less radiation into space; i.e., a warming effect at the surface.

Numerous studies and articles however, including one by Randall (2012, *Atmosphere, Clouds, and Climate*), show that at high altitudes (particularly in the tropics) there is a tendency for the lapse rate to weaken as the surface temperature increases. In other words it doesn't cool as much with increasing altitude. As the surface temperature warms, the lapse rate decreases and the upper troposphere temperature warms. This causes more heat to be emitted into space (warmer bodies emit more heat – infrared radiation), dampening the surface warming trend; i.e., *negative* feedback.

And, according to an article by Ken Gregory (July 2009, *Satellite Finds Warming "Relative" to Humidity*), based on NOAA data from 1948 to 2008, the relative humidity of water vapor in the troposphere declined by 21.5%, especially at higher elevations (over 5 miles high), allowing more and more heat to escape to space. Over the 60 years, the troposphere's negative feedback increased substantially; i.e., it had an increasing cooling effect. But, Gregory says, climate models show relative humidity staying constant which results in positive feedback. It seems that when science reveals results that don't support *the cause*, they are ignored.

There are an immense number of feedbacks operating within the climate system spheres and cycles, and occurring at various time scales. Some feedbacks, such as weathering of rocks (discussed in section 4.6, carbon cycle) occur over millions of years; those involved with ocean circulation (upwelling; down-welling), vegetation (e.g., tree growth and decay), melting massive ice sheets, etc., occur on time scales of centuries. Feedbacks associated with events such as volcanic activity, El Niño, etc., have time scales measured in years, while others, such as the short-term water cycle feedback example presented above occur on time scales of days or weeks. And perhaps the most important feedback network affecting climate is that due to clouds. It is also a source of high uncertainty in climate modeling as expressed by IPCC Assessment Reports (section 4.5). Some of the reasons for such uncertainty in cloud feedback processes include the following:

• Cloud modeling must address microphysical processes involving cloud droplets, ice crystal, and aerosol interactions and feedbacks, with scales as small as millionths of an inch but with model grid size measured in the tens of thousands of square miles
• Clouds change their properties sometimes in minutes; the amount, density, opacity, brightness, size, altitude, etc., which affects the strength and direction of feedbacks; and they are constantly moving horizontally and vertically by wind patterns
• Cloud radiation processes involving the flow of both solar and terrestrial radiation. Clouds reflect incoming solar (shortwave) radiation back into space (cooling), and absorb terrestrial (long wave infrared) radiation; i.e., contribute to the greenhouse effect (warming)
• Clouds also couple climate processes together; they cast shadows on land and oceans, reducing surface temperature and evaporation rates, which affect water cycle feedbacks.

Two relatively simple examples of cloud feedbacks, previously presented in section 4.5 are:

<u>Low Cloud Feedback:</u> Low cumulus clouds are essentially at the same temperature as the Earth surface such that their greenhouse effect is minimal. They can however, be bright at their top which reflects solar radiation; i.e.; negative feedback – cooling

<u>High Cloud Feedback:</u> High, cold cirrus clouds are mostly thin and transparent to incoming solar radiation. They absorb outgoing infrared radiation, contributing to the greenhouse effect; i.e., positive feedback – warming. This feedback process however, can be disturbed by another process, the *Iris Effect* whereby the warming increases evaporation and precipitation, which reduces the amount of cirrus clouds, which reduce their greenhouse effect; i.e., a negative feedback on top of positive feedback, with the net effect, uncertain.

As vital as it is to understand the feedback mechanisms of a system, according to the IPCC AR5 Data Distribution Center (DDC),

"*Other* (sources of uncertainty) *relate to the simulation of various feedback mechanisms in models concerning, <u>for example</u>, water vapour and warming, clouds and radiation, ocean circulation and ice and snow albedo.*"

This admission of uncertainty in feedback mechanisms of such important events further undermines the AGW CO_2 case. Parsing the DDC statement:

- Recall that without a water vapor amplification effect (x 3) there is no case to support the CO_2 hypothesis.
- Clouds cover 67% of the Earth at all times and a 1% variance in cloud cover would wipe out the entire 0.8°C warming associated with CO_2.
- Oceanic circulation includes the global conveyor belt that transports heat around the globe via the oceans and is responsible for the Gulf Stream etc., and

- We saw in section 4.1 that a 3% change in albedo could result in a 3°C difference.

Also, based on the citation's *"for example"* comment, the list (of uncertainties) is not exhaustive.

While it is possible to observe and measure net feedback effects of events such as volcanic eruptions or El Niño, it is not possible to observe the effects of feedbacks, individually or collectively, that take hundreds, thousands, or millions of years to resolve, so these are areas yet to be determined. The present-day *net effect* of the Earth system spheres, cycles, and feedback processes can be measured by observing the difference at the top of the troposphere between incoming solar radiation and outgoing infrared radiation. Section 4.1 shows that the Earth's net radiation is balanced within the margin of error and that Earth is in relative equilibrium.

Section 4.7 Summary: This section provided a macro view of the Earth System, integrating external and internal forcings (discussed in detail in sections 4.1 through 4.6) with the Earth's subsystems or spheres, and the interactivity of spheres with the climate via feedback within numerous cycles of matter and energy, such as the water cycle or carbon cycle. It shows the processes by which the Earth is a stable, self-correcting, adaptable system.

Chapter 5
Models

Global Climate Models (GCM) are computer-based simulations that use mathematical formulae in an attempt to re-create the chemical and physical processes that drive the Earth's climate. They are not used to determine the cause(s) of climate change, since they have already pre-determined that to be human-caused CO_2, but rather to examine the climate response to increasing levels of CO_2.

In science, for a model to be accepted it must be validated, i.e., be able to reproduce known results. GCMs fail to reproduce historical temperatures over the past 100 plus years, including the *impending ice-age* of the 1970s and the relatively stable climate from the late 1990s through about 2014. If simulation models cannot reproduce historical data then they are not validated and should not be relied upon to provide predictions or future scenarios and specifically, they should not be used to formulate energy policies worldwide.

In a study by Antero Ollila, published in the Physical Science International Journal (July, 2017; *Semi Empirical Model of Global Warming Including Cosmic Forces, Greenhouse Gases, and Volcanic Eruptions*), Dr. Ollila concluded that compared to temperature observations, **IPCC model errors are about 49%;** and that's after spending tens of billions of research dollars.

The most important aspect of the models that effects projections is *climate sensitivity*, which is defined as the amount of global surface warming that occurs when the concentration of CO_2 in the atmosphere doubles. If the climate is very sensitive to greenhouse gases, i.e., climate sensitivity is high, then substantial warming can be expected in the coming century. If climate sensitivity is low, then future warming will be lower. The term generally refers to the rise in temperature once the climate system has fully warmed up and stabilized, a process taking over a

thousand years due to the enormous size and heat capacity of the
oceans. This is called equilibrium climate sensitivity (ECS) and is
the most widely used measure. Table 5-1 shows the evolution of
IPCC model estimates for ECS over a nearly 30-year period.

Table 5-1 Evolution of ECR– IPCC Reports

Assessment Reports	ECS Range (°C)
IPCC First Assessment 1990 (FAR)	1.5 – 4.5
IPCC Second Assessment 1995 (SAR)	1.5 – 4.5
IPCC Third Assessment 2001 (TAR)	1.5 – 4.5
IPCC Fourth Assessment 2007 (AR4)	2.0 – 4.5
IPCC Fifth Assessment 2013 (AR5)	1.5 – 4.5

What is startling about the table is that with nearly three
decades of research, massive funding and labor expenditures
(scientists, statisticians, modelers, etc.), enhanced models and
mathematical/statistical methods, and material (satellites, super-
computers etc.), nothing apparently has changed since 1990 that
affects IPCC forecasts. Apparently, that would require another
trip to Mount Sinai!

Each Assessment Report claims a better understanding of: (1)
Radiation – the way in which the input and absorption of solar
radiation by the atmosphere or ocean and the emission of infrared
radiation are handled; (2) Dynamics – the movement of energy
around the globe by winds and ocean currents; (3) Surface
processes –the effects of sea and land ice, snow, vegetation and
the resultant change in albedo, emissivity and surface-
atmosphere energy and moisture interchanges; (4) Chemistry –

the chemical composition of the atmosphere and the interactions with other components (e.g. carbon exchanges between ocean, land and atmosphere); (5) Mathematics - enhanced statistical and parameterization techniques and; (6) Computer power-improved computational power and the resultant better resolution in both time and space. All this progress and IPCC projections have not changed in 30 years.

It is perplexing how the IPCC claims such high confidence in its ECS projections (95%) when, as discussed in various parts of this book, its scientific working group reports are replete with caveats and expressed uncertainties; not only in their understanding of the physical characteristics of the planet and its behavior, but also their ability to mathematically simulate critical processes and interactions without excessive parameterization (Parameterization is discussed below).

Even while recognizing the uncertainties in what are considered to be the most significant contributions to the energy budget (clouds, aerosols, their interactions, feedback, etc.), apart from the (mostly discredited) AR4 report, that increased the bottom of the range for projected ECS from 1.5°C to 2°C, only to be reversed in AR5, it has remained stable. The IPCC maintains its original position that climate sensitivity ranges from 1.5°C to 4.5°C. Decade after decade, the same uncertainties, the same results.

Problems with scale/resolution: The volume of air in the atmosphere is approximately 4 billion cubic miles - Earth's surface, about 200 million square miles, times average atmospheric height of about 20 miles. Figure 5-1 is from an article by Dr. Tim Ball: (2012, *Static Climate Models in a Virtually Unknown Dynamic Atmosphere*). It shows the typical surface and atmosphere grid system used in climate models. Grid spacing of 3° x 3° is an area of approximately 180 x 180 miles or 32,400 square miles.

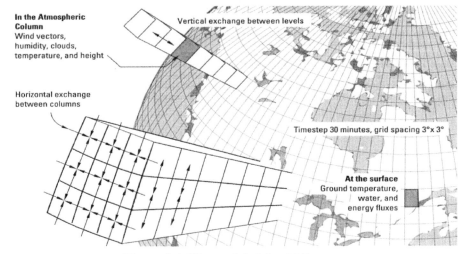

In the Atmospheric Column
Wind vectors, humidity, clouds, temperature, and height

Vertical exchange between levels

Horizontal exchange between columns

Timestep 30 minutes, grid spacing 3°x 3°

At the surface
Ground temperature, water, and energy fluxes

Figure 5-1 Climate Model grid System

The IPCC AR5 Data Distribution Center (DDC) provided grid dimensions employed in its GCMs that are similar to those provided by Dr. Ball's illustration. It depicts the climate using a three-dimensional grid over the globe, typically having a horizontal resolution of between 250 and 600 km (about 155 and 375 miles), 10 to 20 vertical layers for the atmosphere, and up to 30 layers for the oceans. The DDC says,

"...*resolution is thus quite coarse relative to the scale* of exposure units in most impact assessment. Moreover, **many physical processes, such as those related to clouds**, also occur at smaller scales and **cannot be properly modeled.** Instead, their known properties must be averaged over the larger scale in a technique known as **parameterization. This is one source of uncertainty in GCM-based simulations of future climate.**"

And from the AR5 7, Clouds and Aerosols,

"Cloud and aerosol properties vary at scales significantly smaller than those resolved in climate models, and cloud-scale processes respond to aerosol in nuanced ways at these scales. Until sub-grid scale parameterizations of clouds and aerosol– cloud interactions are able to address these issues, model estimates of aerosol–cloud interactions and their radiative effects will carry large uncertainties."

DDC says physical properties of clouds occur at smaller scales (than grid sizes), and IPCC AR5 says clouds and aerosol properties vary at scales significantly smaller than those resolved in climate models. These statements bear some perspective.

Whether horizontal grid sizes are as shown in Figure 5-1 (32,400 square miles) or as stated by the DDC (58,000 square miles), they are large enough to accommodate more than 100 trillion aerosol particles. The average diameter of a cloud droplet is typically 10-15µm (1µm = 1 millionth of a meter); aerosols are about 1/100th the size of a cloud droplet, some as small as one billionth of a meter (0.000,000,039 of an inch). The task, as IPCC scientists say, is daunting.

It follows that since the numerical solution of the model equations only allow the atmospheric state to be known on scales defined by the grid box size, the description of these statistical effects has to be expressed in terms of those scales which as previously stated are in some cases millions or billions of times the size of the particles being modeled. The technique involved is generally referred to as parameterization.

According to Tim Ball (2014, p115),

"parameterization is a fancy word for making up data when it doesn't exist. Because of the lack of data, this occurs for a majority of the surface grids and virtually all the layers above the surface."

Parameterization is an acceptable mathematical technique used to interpolate or approximate data from sparse datasets, data that lies below the resolution of a model, or to simplify complex relationships. However, there is no doubt that extensive use of parameterization and other tuning mechanisms, within and between grids so disproportionate to the physical size and dynamic behavior of the elements being modeled, destroys any semblance of reality. Dr. Ball et al point out that a major failing of the models is that parameterized values from a grid, *augmented* by selective tuning, are then used as true input values in the next grid's computation; i.e., the IPCC et al treat computer models as official data sources when, in fact, computer models should be used to analyze, not create, data.

In a 2007 speech on Climate Change and Development, Antonio Zichichi (Italy's most renowned scientist, professor emeritus of Advanced Physics at the University of Bologna and former president of the World Federation of Scientists, with 10,000 scientists from 115 countries), expressed concern over the amount of parameterization used in GCMs. He quoted John von Nuemann, the great 20th century mathematician, who cautioned his students about employing what he called "free parameters" and is reported to have said,

"If you allow me four free parameters, I can build a mathematical model that describes exactly everything an elephant can do. If you allow me a fifth parameter, the model can make him wiggle his trunk." Such is the power of parameterization and the introduction of variables to curve-fit, i.e., to get the desired result."

Dr. Zichichi also said (2008, *Climate Change, Scientific Consensus & Contrarian Views*),

"Models used by the Intergovernmental Panel on Climate Change (IPCC) are incoherent and invalid from a scientific point of view."

The bottom line is that in such an environment, parameterization allows the output of a grid to be anything analysts desire it to be, so that with deterministic algorithms a modeler can fudge or tune inputs until outputs conform; i.e., they can torture the data until it confesses! IPCC scientists have obvious concerns about the use of such extensive parameterization as can be inferred from the following passages from the IPCC AR5 scientific reports.

WG1, 7.2.3.1 Challenges to Parameterization,

*"The simulation of clouds in modern climate models involves several parameterizations that must work in unison. These include parameterization of turbulence, cumulus convection, microphysical processes, radiative transfer and the resulting cloud amount (including the vertical overlap between different grid levels), as well as sub-grid scale transport of aerosol and chemical species.... **Many cloud processes are unrealistic in current GCMs, and as such their cloud response to climate change remains uncertain.**"*

These statements are incredible given that clouds cover about 67% of the Earth at all times, and remember, according to Dr. John Holdren, President Obama's science advisor for eight years, *"a mere one percent increase in cloud cover would decrease the surface temperature by 0.8°C."*

*"The representation of cloud microphysical processes in climate models is particularly challenging, in part because some of the **fundamental details** of these microphysical processes **are poorly understood** (particularly for ice- and mixed-phase clouds), and because...key*

atmospheric properties occurs at scales significantly smaller than a GCM grid box....Therefore continuing weakness in these parameterizations affects not only modeled climate sensitivity, but also the fidelity with which these other variables can be simulated or projected. "

Note the concern that weakness in parameterization affects climate sensitivity which is the bedrock of the entire debate. And to acknowledge weakness in fidelity (accuracy, precision, reliability and integrity) is incredible.

Furthermore, despite so much expressed uncertainties in understanding the physical processes, inadequate grid resolution, and associated parameterization techniques (what is left?), AR5 increased its confidence in ECS projections from 90% (AR4) to 95%. Imagine making such a presentation to any corporate organization, identifying so many uncertainties in knowledge and processes, yet declaring an increase in confidence and suggesting the corporation should spend additional billions or trillions of dollars continuing on the same course. Incidentally, this is somewhat predictable; research in the psychology of human behavior is replete with examples of overconfidence in uncertain situations when knowledge is low. And recall from the Introduction chapter that one of the effects of *group polarization* is the tendency for a group to become more extreme in its decision-making.

Freeman Dyson, a fellow of the American Physical Society, a fellow of the Royal Society of London, and a member of the U.S. National Academy of Science and, a staunch democrat and President Obama supporter, is a skeptic of the models. In a 2005 Winter Commencement Address at the University of Michigan, and in an interview with Benny Pieser (2007, *The Scientist as Rebel: An Interview with Freeman Dyson*), Dr. Dyson said,

"They **(the models)** *do a poor job of describing clouds, the dust, the chemistry, and the biology of fields, farms, and forests. They* **do not begin to describe the real world…. They are full of fudge factors** *…concerned with processes such as snow melting and vegetation growth that cannot be modeled in detail"* Fudge factors are used, as Dr. Ball said, to *"make up data when it doesn't exist."*

Ten years later, in a 2015 interview, with *The Register*, when asked, "Are climate models getting better?" He responded with,

"I would say the opposite. What has happened in the past 10 years is that the discrepancies between what's observed and what's predicted have become much stronger. **It's clear now the models are wrong,** *but it wasn't so clear 10 years ago."*

Model performance: At no time since its inception in 1988 has the IPCC been able to demonstrate that CO_2 (natural and human-caused combined) is the sole or even dominant driver of global temperature. Its dire predictions are based solely on computer models, models that, in spite of applying favorable (to the IPCC Charter) parameterization techniques have been unsuccessful in replicating historical data, and therefore have not been validated. In laypersons' words, they don't work!

Figure 5-2 shows two versions of IPCC model projections versus observations. The top graph is from the 1990 First Assessment Report (IPCC FAR); the lower graph is from the 2013 Fifth Assessment Report (IPCC AR5), some twenty-three years later.

As can be seen from the figures, the IPCC reduced the slope of its projections (shown as PROJECTED on the figures) from 2.78°C/century (FAR – top figure, in parenthesis) to 1.67°C/century (AR5 – bottom figure, in parenthesis); more than 1°C reduction. This change apparently reflected the stabilization of temperature from late 1990s through 2014. But it didn't affect the protected grail of the ECS range (Table 5-1).

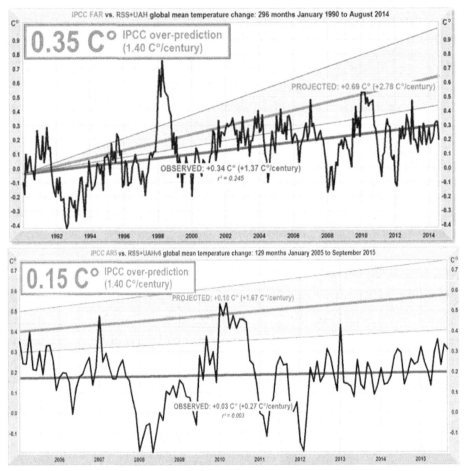

Figure 5-2 IPCC Model Predictions versus Observations

The top figure shows the most probable PROJECTED temperature increasing by 0.69°C over the 296 months, from January 1990 to August 2014, against the OBSERVED temperature increase of 0.34°C over the same period, a difference (over-prediction by IPCC models) of 0.35°C over the period of measurement. Another way of looking at it is that the projected average was almost 100% too high, or too hot: 0.69°C versus

0.35°C over the period of performance, and 2.78°C versus 1.37°C projected for the century.

The lower figure, produced 23 years later, shows the PROJECTED trend over 129 months (from January 2005 to September 2015) to be 0.18°C or 1.67°C per century, against the OBSERVED increase of 0.03°C over the period, or 0.27°C per century, an even worse model fit. The data supports Ollila's (2017) conclusion (model errors are about 49%), i.e., the models fare no better than flipping a coin.

Despite excessive parameterization and tuning the models constantly fail to match observations and, in any other scenario would be discarded. As Richard Feynman said *"If it disagrees with observations – it's wrong. That's all there is to it."*

The following graph (Figure 5-3) was presented by Dr. John Christy, Professor of Atmospheric Science and Director of the Earth System Science Center at the University of Alabama in Huntsville, USA, as part of his testimony before the U.S. House Committee on Natural Resources on May, 2015. It shows the IPCC forecast (dark black line steadily increasing beyond 2020), which is the average of 102 climate models against observations (actual measurements) from two different data collections sources (radiosonde balloons and satellites - circles and squares, respectively) over a 30-year period. No explanation is required other than to point out that only on one occasion, in 1979, did the IPCC model match observed data. As the old adage goes, "even a broken clock is right twice a day."

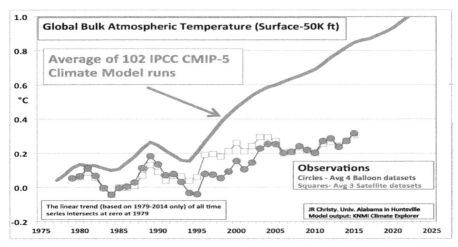

Figure 5-3 Climate Models versus Observations

Roy W. Spencer's blog (October 14th, 2013) provided the graph, Figure 5-4, showing satellite observations of both the surface warming trends (HadCRUT4) and lower atmosphere or troposphere warming trends (UAH) versus IPCC CMIP5 model simulations and predictions (CMIP5: Fifth phase of the Coupled Model Inter-comparison Project was released in AR5 as an improvement that accommodates shortcomings identified in AR4). The average of the models is shown as the dark, unlabeled black line.

Comparing 90 CMIP5 climate model results versus observations (from 1983 to 2013) Dr. Spencer showed that **96.7% of the models warm faster than the observations.** Apparently, the GCMs are parameterized to run hot! And what is most amazing is that the improved CMIP5 models continue to run too hot in hind cast; they still don't replicate historical observations – i.e., they still are not valid. Yet they are used by the IPCC and world governments to drive civilization-changing energy policies.

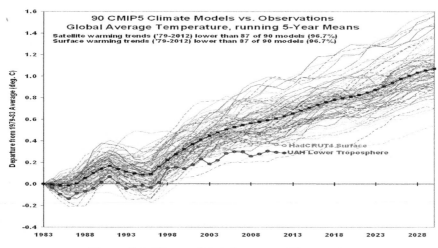

Figure 5-4 Climate Models versus Observations

An interesting observation from Figure 5-4, and a point made by Dr. Robert Brown, a physicist at Duke University, in his June 2013 blog (rgbatduke), is that **the models are supposedly built on top of physics, yet no two of them give anywhere near the same results**. He says,

*"The ensemble of models is completely meaningless, statistically... **an Ouija board would have the same basis of claims** about the future climate as that obtained by averaging the different computational physical models that are subject to all sorts of omitted variable, selected variable, implementation, and initialization bias."*

And in his amazing book (2003, *The Elegant Universe*), physicist Dr Brian Greene says,

"... the ultimate test of a physical theory is its ability to explain and predict physical phenomenon accurately."

Clearly, GCMs fail these criteria.

Chapter 5 Summary: **IPCC GCMs don't work.** Despite excessive use of parameterization techniques and other fudge- factors, adding or deleting parameters, they cannot reproduce observed temperature data – they are not valid, and in any other enterprise would be discarded. Yet, they are used as the sole source to drive multi-billion-dollar energy policies.

This chapter provides examples of quotes from IPCC Assessment Reports which are replete with statements of uncertainty in knowledge and processes in simulating many critical elements.

The IPCC presents its predictions based on GCMs which it portrays as valid mathematical representations of the Earth's systems, interactions, and feedback processes. Yet, its scientific working groups, decade after decade, reveal major uncertainties in both their understanding of the physics and in their ability to model such complexities.

The degree to which the IPCC models apply parameterization techniques, which it acknowledges result in processes that *"are unrealistic in current GCMs"* is most telling in its statement that *"continuing weakness in parameterization affects climate sensitivity and the fidelity with which variables can be simulated or projected."* Climate sensitivity is the bedrock of the AGW argument, and a weakness in fidelity, the degree to which the models represent reality, is a clear admission that GCMs don't work.

This chapter begs the question, "How is it that proponents of the AGW argument, particularly scientists, maintain their position when the models, on which IPCC forecasts and scenarios are based, are not valid?" Is it that they are so invested in the hypothesis that their careers, prestige, their very identity are at stake? If so, and I suspect this to be the case, there may never be a reasonable discourse on the subject. It seriously needs a Martin Luther-type intervention!

Chapter 6
Economics

Clearly, we need a plan to wean ourselves off fossil fuel: eventually it will run out, and once the supply trends toward depletion the world will be in danger. Global alarmists warn us that global warming could lead to war, some even blaming the insurgence of ISIS on global warming. I suggest that running out of fossil fuel without a viable alternative in place is a more likely scenario for conflict. Survival is the basic human motivator, and if nations believe their survival rests on accessing fuel that their neighbor has, that's a dangerous predicament; e.g., Japan in the 1930s and 40s.

The good news is that we have ample time to convert from fossil fuel to some form of alternative energy in an orderly manner, and in a way that doesn't destroy the global economy or quality of life. In September 2006, at the Third OPEC International Seminar, Abdullah S. Jum'ah, President and CEO of Saudi Aramco, said,

"...we are looking at four and a half trillion barrels of potentially recoverable oil. That translates into more than 140 years of supply at today's rate of consumption."

And while the world population has increased significantly since 2006, increasing energy demand, his estimates did not include oil shale. In 2016, the World Energy Council set the total world resources of oil shale equivalent to 6.05 trillion barrels of oil which it says is more than three times the worlds proven oil reserves, estimated to be 1.7 trillion barrels.

In economic theory, the law of supply and demand is considered one of the fundamental principles governing an economy. It is described as the state where as supply increases prices tend to drop or vice versa, and as demand increases prices tend to increase or vice versa.

This simple fact is the main obstacle to eliminating or even severely reducing fossil fuel because supplies are abundant and population (demand) is increasing. More people mean more demand; demand for affordable energy. And the only source of energy that can affordably meet the demand of the world population now and for the foreseeable future is fossil fuel.

According to a 2017 United Nations report (*Economics and Social Affairs*), the current world population of 7.6 billion is expected to reach 8.6 billion in 2030 (15.8% increase), 9.8 billion in 2050 (29% increase over current numbers), and 11.2 billion in 2100, a staggering 47% increase over current numbers. The population increases reflect the amount of additional energy the world will need to produce. China and India (1.4 and 1.3 billion, respectively) comprise 40% of the world population. Neither country will conform to IPCC goals of reducing fossil fuel consumption; China will *review* its position in 2030 when its population is estimated to exceed 1.6 billion. Neither will other *developing* nations. Only the EU, UK, Australia, New Zealand, and the U.S. are participating in CO_2 reduction programs, i.e., about 1.1 billion or 15% of the world population.

According to various sources, these nations collectively emit about 28% of the human part of the CO_2 budget. The rest of the world, those countries not participating in CO_2 reduction programs, emits 72%. To put this in perspective: human contribution to atmospheric CO_2 is about 5% of the total - the total is about 400 ppm, so humans contribute 20 ppm. Those countries committed to reducing their carbon footprint contribute 28% of the 20 ppm, less than 6 ppm; that's 6 parts *per million*! If those nations *eliminated* fossil fuel usage it would reduce atmospheric CO_2 from 400 ppm to 394 ppm. Does anybody believe that reducing atmospheric CO_2 from 400 ppm to 394 ppm would make a difference?

The ratio will obviously become more distorted since among those nations not embarking on *the cause* includes China and India, the two largest and fastest growing populations – the world had better hope skeptics are right!

The following sections address: Energy Supply versus Demand (consumption); Financial Commitments, and then asks the question; Do we need to reduce anthropogenic CO_2 levels? These are followed by a discussion on the Precautionary Principle.

Energy Supply and Demand: The developed world gets 80% of its energy from fossil fuel, which is unsustainable over the long term primarily because (1) easily accessible fossil fuel with eventually run out, and (2) security; apart from the USA, most major exporters of fossil fuel are either in the Middle East, or not-too-friendly, or not-too-stable nations such as Russia, Venezuela, Nigeria, etc.

Dr. David MacKay, in his book, *Sustainable Energy – without the hot air*, provides a unique approach to understanding the problem we face in replacing the energy we get from fossil fuel with renewable alternatives. The main point of his book was to compare the energy that each person receives from fossil fuel to the energy available in various sources of renewable alternatives, using scientifically derived measurements. He reduces the complexities of energy dimensions to units with which we are familiar from our energy bills, i.e., kilowatt-hours, and provides an average kilowatt-hour per day (kWh/d) per person for both supply and demand. He uses Britain as an example of assessing supply and demand but his logic applies to any country; other countries however will have different energy consumption figures and clean energy supply alternatives because of the availability of indigenous resources. (Dr. MacKay's book is available free online from www.withouthotair.com).

Basically, he arrived at a "125 kWh per person, per day" unit of measure for current energy consumption from fossil fuel. This is made up of a per person, per day average use of all significant sources of energy utilization including transportation (cars, planes, and freight), heating and cooling, lighting, information systems and other gadgets, food and manufacturing. For example, driving an average car 50 km per day (approximately 31 miles per day) uses 40 kWh/d, while covering every south-facing roof in Britain with solar-water heating panels would capture 13 kWh/d per person; an annual long-range flight by jet uses 30 kWh/d averaged over the year, and covering the windiest 10% of Britain with onshore wind farms would yield 20 kWh/d per person, etc. He develops a consumption energy baseline and provides six alternative plans that theoretically could supply sufficient clean energy to meet the demand. He doesn't attempt to estimate the cost associated with the plans.

Each of his six plans shows various concoctions of clean energy sources that meet his calculated consumption requirements. Each plan however, requires a greater than 40% reduction in demand from current use; and complex combinations of solar, wind, tide, hydro, bio-fuel, waste incineration, heat pumps, wood, etc., all of which would need to be integrated into the grid. He demonstrates how improved efficiencies could provide the 40% consumption reduction. The more practical plans include nuclear and clean coal energy, to both of which the *Greens* take exception. Three of the plans require importing solar energy from such places as Saudi Arabia which does nothing for the security of supply, one of the main reasons for weaning off fossil fuel. In fact, he emphasizes that any plan not making use of nuclear power or clean coal has to make up the balance from renewable power imported from other countries, the most promising of which he says is from solar power in deserts.

He also acknowledges the environmental impact of his plans. If, for example, it was decided that burning biomass (crops for fuel) was the answer, about 75% of Britain would need to be covered in biomass plantations to meet about 25% of its current electricity demand.

The bottom line, though Dr. MacKay doesn't say so directly, is that for Britain to replace fossil fuel with sustainable energy is not practical. He did say that he fears the maximum Britain would ever get from renewable energy is in the ballpark of 18 kWh/d, per person; significantly lower than the current 125 kWh/d, per person.

And if it's not practical for Britain to extricate itself from fossil fuel, how would the U.S. fare? According to *World Development Indicators*, U.S. per person consumption is approximately 290 kWh/d, i.e., about twice that of Britain. Dr. MacKay estimates that for the U.S. to provide 42 kWh/d, per person, from wind power for example, would require wind farms covering the entire state of California, and that covering the entire state of California with solar panels would provide only about 22% of energy requirements.

Saul Griffith (www.slideshare.net/energyliteracy/longnow-16-jan-09) presents a daunting macro view of the task; not to eliminate CO_2, but to limit it at 450 ppm, about a 14% increase from today's number (around 400 ppm). He asks the question, *"Can there be a global solution for climate change?"* He also provides a numerical evaluation of the energy needed, not as simple to follow as David MacKay's, but it's a global perspective rather than country specific.

In 2009, he calculated that **to limit the CO_2 level at 450 ppm, would require** building about 11.5 new terawatts of clean energy by 2034, i.e., **11.5 trillion watts of new energy**. He says the world currently runs on about 16 terawatts (TW) of energy, most of it burning fossil fuels, and that to level off at 450 ppm the world would have to reduce the fossil fuel burning to 3 TW and produce the rest (13 TW) with renewable energy. At the time of his presentation (2009) about 0.5 TW of energy came from clean hydropower and 1.0 TW from nuclear. That left 11.5 TW to generate from new clean sources.

His analysis shows that to limit carbon dioxide level at 450 ppm, the land area dedicated to renewable energy would occupy a space about the size of Australia. To get to Al Gore and James Hanson's goal of 350 ppm, fossil fuel burning would have to be cut to zero, which means another 3 TW would have to come from renewable sources, expanding the size of land required by an additional 26 percent. There goes New Zealand!

In a 2012 video, Dr. Griffith estimated that to convert 0.8 TW of fossil fuel to clean energy would cost from $2 to $5 trillion. Using those figures to estimate the cost to convert 11.5 TW to clean energy, results in the range of $23 to $57.5 trillion; for zero fossil fuel the range would be from $32 to $90 trillion; this upper limit is greater than the world GDP.

As with Dr. MacKay's assessment, without nuclear, there is not a practical solution to either eliminating anthropogenic CO_2, or even limiting it to 450 ppm with current technology.

Financial Commitments: according to the *CIA's World Factbook*, the World GDP in 2014 was approximately $78 trillion. A 2006 report by Lord Stern, commissioned by the UK government to study the effects of global warming, claimed that an immediate investment of 1% of GDP was required to prevent CO_2 levels exceeding 450 ppm; in 2008 he revised the figure to 2% of GDP per annum as the requirement for achieving stabilization between 500 and 550 ppm. The UK enacted a law, *The Climate Change Act 2008*, which establishes a target of 80% reduction in fossil fuel use by 2050 at a cost of around $560 billion.

The EU, UK, Australia, and New Zealand, with a collective GDP of about $16 trillion are all in, budgeting about 2% per year of their GDP, collectively around $329 billion per year for the foreseeable future. In addition to individual country contributions, the EU Brussels Government (EU has its own Parliament overseeing *sovereign* member nations) has committed to spend 20% of its budget on climate-related projects; for the 7 years from 2014 through 2020 it will spend around $210 billion. This is money it receives from the EU countries and until Brexit, the UK. For further detail of the EU commitment refer to ec.europa.eu/clima/events/0086/funding_en.pdf.

In addition to those amounts, at the 2015 Paris Accord, *developed* countries pledged to provide $100 billion *per year* by 2020 for climate action in developing countries; an insidious redistribution of wealth!

So, what is the impact on those countries embracing *green* energy policies? According to a January 2014 article in The Financial Times,

"High gas and electricity prices will continue to plague Europe for at least 20 years, damaging the competitiveness of industries...."

It continues by stating that because of EU climate change policies,

"Europe will lose a third of its global market share of energy-intensive exports over the next two decades."

Fatih Birol, the International Energy Agency's chief economist, said,

"Europe didn't realize the seriousness of this competitive issue," and that *"European gas import prices are currently around three times higher than in the US while industrial electricity prices are about twice as high."*

Travis Fisher, an economist at the Institute for Energy Research, told The Daily Caller News Foundation,

*"For years, **European countries have forced a transition from reliable, affordable electricity sources like nuclear and coal to unreliable, expensive sources like wind and solar.... The only thing they have to show for it is skyrocketing electricity prices and struggling economies.** In fact, the average German pays three times more for electricity than the average American, due in large part to Germany's top-down energy transition."*

Continuing to describe the financial impact on Germany, Sigmar Gabriel, German Minister for Economic Affairs and Energy from Dec 2013 to Jan 2017, warned that the power price imbalance could cause a *"dramatic de-industrialization"* in the country. Officials in his ministry say no issue worries him more than Germany's *"creeping loss of competitiveness."*

The introductory paragraph in a July 2014 report by Finadvice for the Edison Electric Institute and European clients was,

"Over the last decade, well-intentioned policymakers in Germany and other European countries created renewable energy policies with generous subsidies that have slowly revealed themselves to be unsustainable, resulting in profound, unintended consequences for all industry stakeholders. While these policies have created an impressive roll-out of renewable energy resources, they have also clearly generated disequilibrium in the power markets, resulting in significant increases in energy prices to most users...."

In a 28 November 2015 Special Report, *The Economist* pointed out that,

"Germany has made unusually big mistakes. Handing out enormous long-term subsidies to solar farms was unwise; abolishing nuclear power so quickly is crazy. But Germany's biggest error is one commonly committed by countries that are trying to move away from fossil fuels and towards renewables. It is to ignore the fact that wind and solar power imposes costs on the entire energy system, which go up more than proportionately as they add more."

The average electricity price in European households was 18 cents per kilowatt-hour in 2014 (Germany's 2017 average cost is 35 cents per Kwh), compared to the average price in the U.S. of 10 cents per kilowatt-hour in 2015. And because of its green energy policies, European prices are increasing at a significantly higher rate. Figure 6-1, provided by the US Energy Information Administration July 2017, shows the rate of increase of European prices compared to that of the U.S from 2006 to 2013.

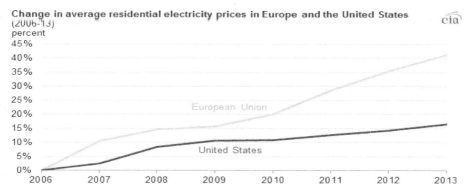

Figure 6-1 Average Residential Energy Prices in Europe and the U.S.A.

While the U.S. is also engaged in subsidizing renewable energy, wind, solar, biofuel, etc., at the federal level it has not lost complete control of its senses and retains fossil and nuclear sources that have enabled it to have more affordable energy. This excludes California, of course, which has an aggressive Renewable Portfolio Standard requiring each energy supplier to procure an increasing amount of energy from renewable sources; 33% by 2020 and 50% by 2030. According to the U.S. Energy Information Administration, California's energy cost in 2017 was 60% higher than the rest of the U.S. and, from 2011 through 2017 the average U.S. energy cost increased by 4% whereas California increased by 23% - insanity!

And, returning to Europe, there is no reason to believe the situation will improve while, because of shale oil and natural gas availability in the U.S., the rate of increase shown in Figure 6-1 for the U.S. should flatten, causing the gap to widen. But, compare these prices, including the U.S., to those countries which have not embraced *green energy* policies – China is around 5 cents/Kwh, India around 7 cents, and Russia averages 8 cents per Kwh.

Incidentally, France, which derives about 75% of its electricity from nuclear energy, is about 26.5% cheaper than the EU average, and about 45% cheaper than in Germany. Because of its low cost, about 12 cents per Kwh, France is the world's largest net exporter of electricity gaining over 3.5 billion dollars per year.

In addition to the financial impacts of its policies, Germany has also traded national security for *the cause*. It relies on foreign countries to make up its energy shortfalls; one of these being Russia against whom, ironically, Germany supports sanctions since its annexation of Crimea in 2014 - about 35% of Germany's gas is imported from Russia.

So, high energy costs, reduced trade competitiveness, billions of dollars annually to convert to green energy, and loss of national security to combat an enemy that has not been proven to exist. The next question then is "do we need to?" The following section addresses this question.

Do we need to reduce anthropogenic CO_2 levels? We saw earlier in the section on Greenhouses gases (Chapter 4, Section 4.6) that the Kyoto Protocol targeted reduction in human-caused greenhouse gases was miniscule – a reduction of a few hundredths of a percent by 2020. The following analysis illustrates the effect of human-caused CO_2 on temperature:

Greenhouse effect = +33 °C (i.e., without the greenhouse effect we would be 33 degrees colder)
Water vapor causes 95% of the effect = 31.35°C
Atmospheric CO_2 contributes about 5% of the effect = 1.65°C
Human-caused CO_2 is about 5% of atmospheric CO_2 = 0.83°C

i.e., shutting down the world economy and spending hundreds of billion dollars per year would reduce the temperature due to the anthropogenic CO_2 effect by less than 0.1 °C, not even a noticeable nor measurable amount using standard thermometers.

The astronomical cost to achieve so little should be sufficient to deter an overzealous approach to weaning the world off carbon fuel, but of course there's too much at stake for prominent alarmists to concede to rationality. So, based on the calculations shown above we clearly don't need to expedite a transition from fossil fuel anytime soon, but should we?

Not everyone agrees that *any* approach to reduce fossil fuel is a good idea. At a 2000 conference in New Delhi, Dr. Deepak Lal, Professor of International Development Studies at the University of California, at Los Angeles, talked of the new imperialist threat posed by the ecological movement, particularly for the developing countries. Prof. Lal warned that the Green movement is a modern secular religious movement engaged in a world-wide crusade that needs to be fiercely resisted, calling the movement the *"new cultural imperialism."* He says,

*"Globally, poverty is the number one cause of preventable illness and premature death. Developing countries require affordable, scalable energy to lift their peoples out of poverty... **The greatest threat to the alleviation of the structural poverty of the Third World is the***

continuing campaign by western governments, egged on by some climate scientists and green activists, to curb greenhouse gas emissions, primarily the CO_2 from burning fossil fuels."

Prof. Lal called on India and other developing countries to stand up to what he calls an insidious threat coming from the *global greens*. He called for India to consider withdrawing from a range of international environmental agreements and conventions, including the Kyoto Protocol, claiming that many of the implications of these agreements would impose a very heavy burden, particularly on the poor. It is easy to understand Professor Lal's position: according to the World Bank Report (2013), more than 300 million people in India are without electricity.

Dr. Bjorn Lomborg (2010, *Cool It*) argues that the cost-benefit trade-off of the trillions of dollars to accomplish miniscule CO_2 and temperature reduction presents a moral and/or ethical dilemma. He asks the question, what is the greatest good for humans – direct help now and save millions of lives or strive for a slightly lower, almost imperceptible temperature decrease in 2100? He expresses a point shared by Professor Lal et al that there is an immediate need among poor nations and that money would be better spent on relieving existing sickness and poverty.

According to World Energy Outlook (*WEO-2016*), an estimated **1.2 billion people did not have access to electricity** and more than 2.7 billion people are estimated to have relied on the traditional use of solid biomass for cooking, typically using inefficient stoves or open fires in poorly ventilated spaces. In a special report on *Energy and Air Pollution* (July 2016) under the World Energy Outlook, the International Energy Agency stated that an estimated **3.5 million deaths per year are connected with energy poverty** and the use of biomass and kerosene for cooking and lighting. Surely, Professors Lomborg and Lal have a strong argument that the unwarranted concentration on reducing CO_2

emissions is deflecting funds and attention from these more immediate and worthwhile projects.

Not only are AGW activists siphoning massive amounts of money from more needy projects, studies have shown that production of one of their favorite alternative fuels, **ethanol, increases the carbon footprint**. A 2008 study by Timothy Searchinger et al (science.sciencemag.org/content), found that corn-based ethanol, instead of producing a 20% savings, nearly doubles greenhouse emissions over 30 years. Biofuels from switch grass, if grown on U.S. corn lands, increase emissions by 50%. This is due to the carbon emissions that occur as farmers worldwide respond to higher prices and convert forest and grassland to new cropland to replace the grain to produce biofuels. It was in 2008 that this information was first made public, and since then numerous studies support the theory that **biofuel production of a gallon of ethanol increases the carbon footprint more than that of a gallon of gasoline**. However, since science cannot get in the way of ideology or billions of dollars profit, converting agricultural land for biofuel continues to be subsidized by the US, EU, UK, Australia, and New Zealand.

Over the past several decades a concept called the Precautionary Principle has insidiously become a standard in international law and treaties, and is more and more becoming a default policy for nations, states, and regions as a decision making tool that can be applied without the need for scientific evidence or traditional economic analyses. The following section discusses the principle and its use by government agencies, and the IPCC.

Precautionary Principle (PP): While there is no single definition of the precautionary principle, all versions seem to have as their genesis the Wingspread Declaration, first published in 1988,

"When an activity raises threats of harm to human health or the environment, precautionary measures should be taken even if some cause and effect relationships are not established scientifically."

Unfortunately, in the hands of governments this is tantamount to winning the lottery; they can, with a little tweaking to the Wingspread Declaration version, introduce ideology into their decision-making with reduced scrutiny or the need for due diligence. Consider the following adaptation by the *Center for Health, Environment, and Justice* where it basically eliminates traditional economic analyses,

"Precaution is a systemic change that transforms the way we approach environmental regulation and decision making. This change is rooted in a paradigm shift away from risk/benefit and cost/benefit decision-making that asks, "What level of harm is acceptable?" to a precautionary approach which asks, "How can we prevent harm?" (Maria Mergel, May 11, 2016, *Center for Health, Environment, and Justice*).

Hmm, well we could start *preventing harm* by reducing the speed limit for automobiles to say 5 mph. According to the *Association for Safe International Road Travel* (2012-2017), there are about 1.3 million driving-related deaths per year globally of which around 30% were attributed to speeding.

There are claims that applying the PP in certain circumstances makes sense, such as placing the onus of proof of safety on suppliers of new drugs, chemicals, technology, etc., so that the provider of a product must prove that its product does not harm human health prior to releasing it to the public; in essence, the product is assumed to be harmful until proven by the supplier that it is not. The same results would be achieved by applying the scientific method: the supplier's hypothesis is that the product is safe; the null hypothesis to be tested (and that must be rejected) is that it is not safe. However, as illustrated by the Center for Health,

Environment and Justice's version of the PP, it is an opportunity for governments to by-pass rational decision-making in favor of ideology, as is the case when applied to climate change.

One of the most famous statements of the precautionary principle is Principle 15 of the 1992 Rio Declaration of the U.N. Conference on Environment and Development, and which is adopted by the IPCC to support its charter,

"In order to protect the environment, the precautionary approach shall be widely applied by States according to their capabilities. Where there are threats of serious or irreversible damage, lack of full scientific certainty shall not be used as a reason for postponing cost-effective measures to prevent environmental degradation."

Note the changes from the Wingspread version: instead of <u>some</u> *cause-and-effect relationships are not established scientifically*, we now have lack of FULL scientific certainty as the benchmark. **P.S., all scientific theories lack full certainty!**

At a glance this IPCC precautionary principle version may seem reasonable and perhaps could even be considered to be plain common sense, but upon close examination, however, applying this principle to world energy policy making, is fraught with the potential for abuse, a license to do whatever they want; primarily in the interpretation of the Principle and subsequent decision-making. Who decides what a State's capability is? Who decides what are (real) threats and their level of seriousness? Who decides how much *"lack of full scientific certainty"* is acceptable? Who decides what *"cost-effective measures"* are? And who decides the degree of degradation that is to be prevented? If the answer to these questions is "the government" then we have reason to be concerned. If the answer is the United Nations through its IPCC proxy, then we reason to be *very* concerned.   To quote Ronald

Reagan: *"The most terrifying words in the English language are: I'm from the government and I'm here to help."*

Also note the alleged objective to *"prevent environmental degradation."* What rational person doesn't want to protect the environment? It's just another straw man distraction; anyone objecting to the rest of the statement is against protecting the environment. Furthermore, as we saw in Chapter 1, according to its former chairman Edenhofer, IPCC international climate policy is not about saving the environment, it's about redistributing wealth.

In any case, the irony of applying the PP to the IPCC charter is inescapable: the precautionary principle is based on acknowledgement that science is an active knowledge system in which new discoveries are constantly made, and that scientific evidence is always incomplete and uncertain. So, **when the IPCC is making policy, the science is *not* settled, but when its advocates are manipulating the public, the science *is* settled.** But to repeat the words of global environmental activist Paul Watson,

"It doesn't matter what is true; it only matters what people believe is true."

And, when there does seem to be a reasonable case for applying the precautionary principle, the IPCC does nothing. In its Fifth Assessment Report (2013), the IPCC noted that,

"Increasing demand for biofuels shifts land from food to fuel production, which may increase food prices disproportionally affecting the poor... Despite high agreement that biofuel production plays a role in food prices, little consensus exists on the size of this influence."

First, there is no *"may"* cause food prices to increase; it does! This, together with scientific evidence that biofuel production increases the carbon footprint, certainly falls into the *why not* category. Perhaps in this case, ideology is runner-up to special interests.

<u>Chapter 6 Summary:</u> **There is no alternative to fossil fuel in the foreseeable future**. The world population is exploding and fossil fuel is the only affordable and practical source of energy that will meet demand for the foreseeable future. Supply and demand will ensure the continuing use of fossil fuel for most of the world's nations so long as it is available and cost-effective. Because of the massive resources available we have generations in which to develop and implement alternative energy sources.

We saw that not only is it impossible to achieve IPCC targets, it is not necessary; eliminating human-caused CO_2 would reduce average global temperatures by less than 0.1°C. Furthermore, *the cause* creates significant environmental impacts and consumes massive funds that could be used to alleviate poverty, sickness, suffering, and death for millions of people around the world.

Ironically, it has been shown that biofuel production, the answer to the prayers of many green activists, in addition to raising food prices, increases the carbon footprint.

Until a new, realistic alternative energy source is found we will use fossil fuel. The relatively few nations that have embraced an aggressive green policy are doing so at the expense of their citizens' quality of life, industrial competitiveness, and potentially national security.

Finally, the IPCC implicitly acknowledges that the science is not settled by invoking the precautionary principle which, by definition, is only enacted in the absence of scientific certainty.

Chapter 7
Discussion

The Earth has been on a slow warming trend since the end of the Little Ice Age (LIA), around the mid-19th century. This is a good thing; according to Plimer (2009), the LIA was characterized by crop failure, famine and disease, war, social disruption and depopulation, and even hungry wandering climate refugees resorting to cannibalism and, remember from Chapter 1 the most active period of witchcraft trials in Europe coincided with the 400-year Little Ice Age when witches were blamed for poor harvests. Furthermore, we will see later in this chapter that cold climates are responsible for many more deaths than warm climates.

This discussion begins with a set of questions and answers that reveal what I have learned from my research over the past two years, followed by how I perceive AGW alarmists to be *informally* organized, and a suggested way to move forward.

Do I believe the climate is warming? Yes.

Do I know how much it has warmed over the past century or so? No, pre-satellite records (1979) are suspect and in some cases manipulated to support *the cause*, but I accept the most often quoted range of 0.6 to 0.8°C for this discussion.

Do I believe CO_2 has increased over the past century or so? Yes.

Do I know how much CO_2 has increased over the past century or so? No, my research indicates that early chemical-based measurements, prior to Mauna Loa's infrared sensor method, were cherry-picked to support *the cause*, but I accept an increase of about 120 ppm (from 280 to 400 ppm) for this discussion.

Do I believe human-caused CO_2 is the reason for the increase in temperature? No. Temperatures have oscillated from warnings of an impending ice age in the 1970s to warnings of uninhabitable warm temperatures in the 1980s and 1990s, to no change, *a pause*, for more than 15 years from 1998 through 2014, while CO_2 steadily increased. Arguably the warmest year of the 20th century was 1934 when CO_2 levels were significantly lower than present amounts, possibly as low as 300 ppm, well below IPCC et al targets, almost at the generally accepted pre-industrial level (280 ppm). The relationship between CO_2 and temperature is a minor factor within a complex Earth system, whose effect on climate is dwarfed by natural events, complex interactions, and feedback mechanisms. CO_2 does not control climate; it *contributes* to climate.

Are you certain that anthropogenic CO_2 is not causing global warming? No. Despite the *science is settled* nonsense, science doesn't deal in *certainties*, but when a hypothesis fails a (single) test it is discarded. The claim that CO_2 is the dominant cause of climate change or global warming, fails many tests as indicated in the previous Q&A.

CO_2 is a greenhouse gas (GHG), and GHGs impact global temperature, correct? Yes. CO_2 is a GHG and GHGs do impact global temperature which, incidentally, allows for a habitable planet, but CO_2 is a minor greenhouse gas (approximately 5% of total) compared to water vapor (95%) which is a much more effective greenhouse gas. In fact, **without the assumption** by IPCC et al **that water vapor** *amplifies* **the CO_2 signal** *by a factor of three*, **there is no case for CO_2 being a significant contributor to climate change or global warming**. This is not controversial even among AGW advocate *scientists* and, in addition to numerous scientific studies that suggest this amplification factor is too high, **NASA-funded research shows that the models exaggerate the degree of amplification**.

Of the 5% CO_2 amount (of GHGs), about 5% is human-caused, so **only 0.25%** (5 x 0.05) **of the total greenhouse gas effect is due to humans,** primarily from burning fossil fuel; **more than 99% of GHGs are from natural sources** (evaporation, decomposition, respiration, volcanoes, etc.).

But what if CO_2 continues to increase?   Even if it does, its greenhouse absorption capability (the amount of infrared it can absorb) is nearly saturated so more CO_2 won't be significantly additive and, since the greenhouse effect of CO_2 is approximately logarithmic, and about 70% of the effects has already occurred, additional quantities will have diminishing effects.

Furthermore, from a system point of view it is most likely that the Earth is adaptive to its environment and through negative (stabilizing) feedback, maintains relative equilibrium; i.e., it is self-correcting. For example, we saw in Chapter 4, section 4.6, that adding CO_2 to the atmosphere increased the effectiveness of photosynthesis by generating larger, more CO_2 - absorbing plants, a natural adaptation and stabilizing negative feedback. And Randall (2012, *Atmosphere, Clouds, and Climate*) says the Earth's water cycle adjusts to perturbations to its equilibrium, "*quite naturally*" through negative feedback.

What about the 97% consensus that says the science is settled and human-caused CO_2 is responsible for global warming (and cooling apparently!)? **Consensus is a non-scientific *talking point* invented by politicians to silence opposition and avoid serious debate**.

Do I know the cause of the global warming and climate change? Global warming, the average temperature increase over the past hundred years or so, no; and neither do anyone else. As the IPCC et al constantly remind us there are too many uncertainties in critical climate elements particularly with clouds and aerosols and remember, clouds cover around 67% of the Earth at all times. Many scientists, including President Obama's Science Advisor, Dr. John Holdren acknowledge that **a 1% change in cloudiness could account for all the 20th century warming**.

Climate change however, versus global warming, is an evolutionary process caused in part by those elements discussed in Chapter 4: the Sun's changing activity; orbital and attitudinal changes of the Earth relative to the Sun; the continuous movement of continents; atmospheric and oceanic circulation; oceanic oscillations; the hydrological and carbon cycles, etc., and most importantly, the complex interactions and feedback processes within and among those components.

If you believe it is warming, and you don't know the cause, but others say the cause is human CO_2 primarily from burning fossil fuel, wouldn't the precautionary principle be appropriate; i.e., why not abandon fossil fuel, just in case? If the cost was small, and the impact on society was insignificant I may agree, but neither of those is real. The cost is prohibitive, running into trillions of dollars (that we don't have) and, as Lomborg (2009, *Cool It*) reminds us, the impact on poor, developing nations is nothing short of genocide with millions of premature deaths resulting from the lack of affordable energy. But also, it would be conceding to pseudoscience and worse, to an ideology that seeks to destroy western society; CO_2 is not the cause of climate change, it is simply another contributor within a complex Earth system.

Do I believe in the efficacy of IPCC climate forecasts and scenarios? No. And many eminent scientists from both sides of the political spectrum, including Dr. Freeman Dyson, a life-long democrat and Obama supporter, agree that the global climate models are useless. Dr. Dyson says the *"models do not begin to describe the real world"* and that *"they are full of fudge-factors."* Dr. Robert Brown notes that while the various models are supposedly based on the laws of physics, *"no two of them give anywhere near the same result."* And physicist, Dr Brian Greene (2003, *The Elegant Universe*) says *"The ultimate test of a physical theory is its ability to explain and predict physical phenomenon accurately."* Chapter 5 shows that they cannot reproduce (explain) physical phenomenon (past events) nor accurately predict physical phenomenon (Figures 5-2, 5-3, and 5-4), so how can they be relied on to provide believable forecasts for the next 100 years? It's astonishing that governments are making such revolutionary policy changes based on models that cannot be validated.

If the science is sufficient to convince you that CO_2 is not the cause of climate change, then obviously others are aware of the same science, so why don't they arrive at the same conclusion? Many do. But this is not about science, it's about ideology, political aspirations, survival (careers etc.), in-group acceptance, etc, discussed below. Science has become irrelevant in this so-called debate. Chapter 1 provides quotes from leaders of *the cause* clearly demonstrating that *environmentalism* is a mere proxy used to coalesce the masses. This was enunciated by Ottmar Edenhofer, IPCC chair from 2008 to 2015 who, in a 2010 interview, unmasked the environmental myth, or as Edenhofer called it, *the illusion,* saying,

"One has to free oneself from the illusion that international climate policy is environmental policy.... We redistribute de facto the world's wealth by climate policy."

The long-term goal of the ideologues, to create a world socialist government, under the auspices of the United Nations that controls world finances through draconian energy policies, wouldn't be as popular a message as *Saving the Planet*.

<u>What else should readers know</u>? Readers should be aware of:

• **The scandals and deceptions by politicians, media, scientists, and government entities,** including those discussed in Chapter 2, **to advance a cause that cannot be advanced through science**.

•   **Claims of** *consensus*, a non-scientific card that is played to silence dissent without discourse. It is part of demagogue's Orwellian *Newspeak* lexicon along with *denier*, etc.

• The IPCC and government agencies use of **the Precautionary Principle**, a decision-making process that *is used only in the absence of settled science*; is a means of by-passing rational decision-making processes.

• **If the science is settled why continue to spend billions of dollars each year on climate research?** But of course, the science is not settled, even by its own admissions.

•   IPCC authors and scientists acknowledge decade after decade with each Assessment Report that clouds and aerosols (two major sources of climate forcing) continue to be major sources of uncertainty. They do not fully understand their physical behavior or their interactivity and feedback processes. See the numerous IPCC author statements of uncertainty in Chapter 4, section 4.5.

• The most important aspect of IPCC models is projecting scenarios based on climate sensitivity, the amount of global warming that occurs when CO_2 is doubled. After 30 years of research its estimates have not changed; the results appear to be hard-wired into the simulations.

• IPCC statisticians and modelers express uncertainty in their mathematical *parameterization* techniques (called *fudge-factors* by many scientists) they use to fill in major data gaps, extrapolate, and interpolate data, in their models. And they treat model output as if it were experimental data. Models are not evidence; they are a tool for examining data and providing *what-if* scenarios.

• Scientists do not understand the causes of oceanic oscillations, the major sources for transporting heat and humidity around the planet; nobody knows what causes hurricanes.

• The carbon cycle plays a major role in maintaining the Earth's climate equilibrium yet, the U.S. DOE expresses uncertainty in the biological part of the carbon cycle and NASA expresses uncertainty in its understanding of the role of oceans in the carbon cycle - and   oceans cover more than 70% of the Earth's surface. Then there is the case of the *missing sink*. For more than three decades scientists studying the global carbon cycle have identified an imbalance in the carbon budget; measurements show that only about three-quarters of the CO_2 being produced, natural and human-caused, was accumulating in the atmosphere and oceans. The remainder, about 25%, is missing! No one *knows* where it is going.

• The AGW case (IPCC) is based to a large degree on human emissions remaining in the atmosphere for 50-200 years. Numerous independent studies assert the incubation period is more in the region of 5-10 years, which destroys the AGW claim.

• The IPCC ignores the continuous out-gassing of CO_2 from earthquakes, of which there are more than 10,000 a year, and sub-surface volcanoes, which comprise about 85% of all volcanoes.

It is paradoxical how the IPCC expresses such high confidence in its forecasts (95%) when, at the same time, acknowledges it is *highly uncertain* about major climate influences. This is however, classical predictable behavior; in uncertain situations when knowledge is low, people tend to be overconfident, and this bias is exacerbated in group settings where *group polarization* leads to more extreme positions.

Very little attention is paid to the urban heat island (UHI) effect whereby temperatures in cities (where many measurements are recorded) increase dramatically with increasing population. For example, according to Lomborg (2009, *Cool It*), satellite measurements of Houston, over the period 1990 to 2000, during which time the city population increased by around 300,000 (a 20 % increase), the nighttime surface temperature increased about 0.8°C; i.e., the same increase that is attributed to the entire CO_2 increase over the past century or so. And Spencer (2017, *Climate Change, The Facts*) shows that station warming bias is very sensitive to even small population increases. He shows a 0.6°C increase in UHI effect between a completely rural (people free) environment to one with a population of only 1,000 per km^2. With the soaring rate of population increase (discussed in Chapter 6), it is reasonable to assume temperature *measurements* will continue to increase regardless of actual global temperatures.

Readers should also be aware of misleading *partial* statistical data; i.e., *half-truths,* and also straw man claims, particularly those made with *conditional* statements. In his AIT book, Al Gore states that, *"In the summer of 2003, Europe was hit by a massive heat wave that killed 35,000 people."*

Half-truths: While his claim was based on factual data, around 35,000 European deaths in the summer of 2003 were recorded as heat-related, he (conveniently) left the reader with an incomplete picture.

It is well documented that *cold-related* deaths are about 10 times the number of heat-related deaths. In a 2015 research article published in The Lancet, funded by the UK Medical Council, Antonio Gasparrini PhD, et al (*Mortality risk attributable to high and low ambient temperature: a multicounty observational study*), analyzed 74 million deaths that occurred from 1985 to 2012, collecting data from 384 locations in Australia, Brazil, Canada, China, Italy, Japan, South Korea, Spain, Sweden, Taiwan, Thailand, UK, and the USA. As shown in Figure 7-1, they found that **most of the temperature-related mortality was attributable by a significant margin to cold temperatures**.

Note it wasn't *extreme* cold temperatures that had the biggest impact on fatalities, but moderately cold temperatures. As Huxley said in Brave New World,

"Great is truth, but still greater, from a practical point of view, is silence about truth."

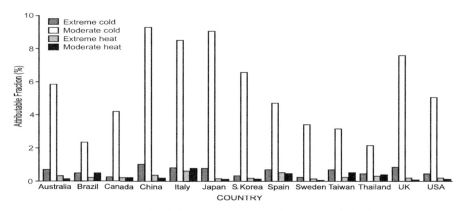

Figure 7-1: Cold versus Heat-Related Deaths (%)

Conditional statements: An example of a *conditional* straw man misrepresentation is Gore's claim that,

"If Greenland melted or broke up and slipped into the sea – or if half of Greenland and half of Antarctica melted or broke up and slipped into the sea, sea levels worldwide would increase by between 18 and 20 feet."

In his movie and book, he says that if this occurred Miami would be underwater, Amsterdam would vanish, in Bangladesh and Calcutta 60 million people would be uprooted, and the site of the World Trade Center Memorial would be underwater. This scenario might very well be true *if* sea levels rise by 18 to 20 feet, but it is a fairy tale and has no basis of fact. If the Gulf Stream stopped, Europe would freeze; if El Niño started to occur every year, we would be in the grip of a global disaster, if pigs could fly!

Incidentally, the IPCC estimates *one*-foot sea-level increase this century, the same as for the last century. Gore's 18 to 20 feet is fiction and most likely intended to mislead the public but he dissembled the story by starting with the conditional "*if.*"

Another misleading part of Gore's book and movie is showing plumes of smoke belching from chimney stacks, implying it is human-caused CO_2 emissions, when CO_2 is a colorless gas.

What if skeptics are right? If we focus solely on reducing anthropogenic CO_2, and the climate we are experiencing are part of a natural climate cycle, then we are not preparing adequately for either the eventuality of continued global warming due to natural causes or what could be the next Ice Age. In fact, in the latter scenario we are preparing entirely the wrong way by reducing CO_2 when, if cooling occurs, we will need more greenhouse protection.

The next section provides my view of a construct comprising an *informal* organizational pyramid or hierarchy that maintains the AGW advocacy, as follows:

At the top are the ideologues: they have a long-term vision, an overarching objective, of establishing a World Socialist Government. The original protagonists, Maurice Strong, Founding Executive Director of the UN Environment Program (UNEP), et al, obviously considered that they could achieve their political objectives by controlling world energy, disguised under a banner of concern for the environment; every country needs energy and, every country has environmental issues and concerns. But as shown earlier the *environment deception* became clear in 2010 when Ottmar Edenhofer, IPCC chair from 2008 to 2015 said that **to believe it's about the environment is an** *illusion.*

On a long journey that began in the 1960s with the Club of Rome (which considered that the industrialized population would outgrow all resources – an expansion of the Malthus theory), they advanced their cause by embracing *environmentalism*, a concern at that time held by most of the general public because of nuclear testing and the global pollution of air and water. Emboldened by their success in 1987 with the Montreal Protocol to *save the planet* by reducing or eliminating carbon fluorocarbon chemicals (which *were* shown in laboratories to destroy ozone) they contributed to the formation of the IPCC in 1988, followed in 1992 by an extravaganza event in Rio de Janeiro, Brazil. It was conducted by the United Nations Conference on Environment and Development (UNCED), labeled *The Earth Summit*. The Conference General Secretary was Maurice Strong. Since then, most western governments have bought into the dogma that humans are destroying the planet. For ideologues of any cause, *the end justifies the means*. Chapters 1 and 2 provide numerous examples of deceit, lies, and threats used to advance *the cause* and achieve the end objective, a world socialist government.

At the next level of the pyramid are politicians. Some may share the ideology of the top level, but most likely they are opportunists carrying the water for the ideologues. Their main objectives are more to do with prestige, wealth, and power, and according to Psychologist Dr. Leon F Seltzer (2011, *Evolution of the Self: Narcissism: Why it is so rampant in Politics*), the more the better. They are the only species with the hubris to have self-promoting rallies to raise money so that they can keep their job! Since level-one ideologues have convinced most western governments to provide billions of dollars to *the cause*, it is fertile ground for politicians on which to feast. Al Gore is the poster child but is accompanied in his zeal by numerous other politicians worldwide who never miss an opportunity for mass media exposure and particularly if it *coincides* with a *feel-good* cause. These are the foot soldiers for the ideologues, they demagogue the public with propaganda and half-truths; their existence is based on a divisive agenda; they are generally skilled orators, such as Gore, who can stir emotions in the public psyche as he did with his AIT movie. They also don't seem to be inhibited by a need for truth; again, *the end justifies the means*. But this time, the end is more self-aggrandizing than ideological.

Media occupies the next level; it, like politicians is driven by self-interest. As discussed in Chapter 1, without sensational headlines media would be ignored. Survival (profitability) is based on readership/viewership and it knows that without alarmism there is little public interest. And since the public's attention span can be measured in days, sequels must occur regularly and each one must be more sensational. Just as politicians carry the water for the ideologues, media carry the water for politicians; they attempt to *legitimize* the political propaganda. As with politicians who repeat slogans and *talking points* over and over again, media repeats stories, generally with increasing hype, knowing that people are highly influenced by the *illusory truth effect*, a cognitive bias whereby people equate repetition with truth; the illusory truth effect is employed widely by politicians and media to manipulate the public.

Scientists occupy the next level. While some may be driven by altruism for a more verdant planet, because of the success of levels 1 and 2, scientists benefit by being compliant, not rocking the global-warming boat, in securing research grants, promotion opportunities, and perhaps just as important not being exposed to hostility, job loss, etc. Chapter 2 demonstrates the extent to which scientists will stray from scientific integrity and ethical standards to ensure their safety (career, opportunity, etc) and acceptance in their communities, *in-groups*. As Gore so eloquently puts it, *"It is difficult for a man to understand something, if his salary depends on his not understanding it."*

At the bottom of the pyramid are laypeople. Here, I believe tribalism prevails. There is in general no other reason to have a strong position other than to be on the same side, in the same tribe or group, as those with whom a person generally associates and agrees with on other important topics, such as political affiliation. According to a poll by Pew Research Center, June 2015, about 64% of democrats believe global warming is caused by humans while only 22% of republicans agree. Interestingly, when combined with another tribal affiliation, religion, the poll showed that the political associations (tribes) are dominant - 62% of democrat Catholics attribute global warming to humans while 24% of republican Catholics agree, indicating that even with Pope Francis being an ardent, outspoken AGW advocate his views apparently have no (statistical) effect (margin of error was +/-3.5%); political tribes are very strongly bonded.

Those who have read George Orwell's book (*1984*), and recall Chapters 1 and 2 of this book, may at this stage be experiencing a *déjà vu* moment.

Oceania – the IPCC and major western governments ruled by small groups operating behind the scenes; the *Inner Party* (ideologues; the self-appointed representatives of *Big Brother*). They make policies (Precautionary Principle; Kyoto Accord; CO_2 Targets, and; even enact Laws as in the UK, California, et al), they select winners and losers with massive federal funds (wind, solar power, ethanol subsidies, research grants, etc.), promotions, acceptance, etc. awarded to *winners;* contempt, threats, public shaming and career destruction for *losers.* And they develop strategies to control the thoughts of the masses, the *Proles.* For example, the BBC has banned skeptics from participating in climate discussions so that viewers are subjected only to the *Party* line and recently (October 2018), DeBlasio, Mayor of New York City, approved the display of brightly lit signs that read *Climate Denial Kills.* Big Brother is watching!

Ingsoc – the ruling philosophy requiring complete obedience. It is built on a vast system of psychological control: group behavior described in the Introduction; Chapter 1 - vocabulary, mind control, fear, media sensationalism, association with evil doers, consensus, appeal to authority, etc., whereby any type of free thought (skepticism) is subject to punishment; loss of career or growth opportunities, harassment, shunning, etc., even calls to murder dissenters.

The Ministry of Truth – a propaganda machine that revises history to make the illusion appear true. Chapter 2 – IPCC changing scientific reports; Articles of Deception – cherry-picking data time-frames; NASA changing historical data and graphs, and denying access to information that may challenge Ingsoc, *the cause*; the Hockey-Stick ruse – which removed the Middle-Ages Warming period; Climategate – which showed a conspiracy to conceal and mislead; manipulating a Children's Book to brainwash pliable minds; the movie, *An Inconvenient Truth*, that contains between nine and 35 scientific errors (all in the direction of *the cause* – a 1 in 34 million chance!) being shown to children in UK public schools (presumably for their *re-education*).

Newspeak – a language designed to shape thought; *the science is settled, consensus, denier*, etc.

Thought Police - *Thinkpol* punishes personal and political thoughts unapproved by the *Party*. Skeptics are maligned by *Party* members and their career opportunities are extinguished should their thoughts be made known. Dr. Curry is a climatologist and former chair of the School of Earth and Atmospheric Sciences at the Georgia Institute of Technology. She has a PhD in geophysical sciences. She was called a *heretic* by colleagues and a *denier* by members of Congress, and intimidated by a House Representative sending a letter to the President of her University. She was guilty of *thought-crime*, the *criminal* act of holding beliefs or doubts that oppose or question Ingsoc, the ruling philosophy, *the cause*, that anthropogenic CO_2 causes global warming. *The State demands absolute submission!*

How do you suggest we should proceed? David Hume said "*A wise man proportions his belief to the evidence.*" I propose the following approach as a basis to move toward a problem-solving direction:

1) Separate the issues of human emissions and global warming from the environmental impact of fossil fuel use. Sensible skeptics of the AGW position accept that fossil fuel extraction and burning are major sources of environmental and public health concern.
2) Accept that the Earth is warming – it is.
3) Accept that atmospheric CO_2 is increasing – it is.
4) Accept that humans are contributing to increased CO_2 levels – we are.
5) Accept that global warming could be the result of anthropogenic CO_2, or it could be caused by natural events – it doesn't matter because,
6) Most of the world will not comply with IPCC et al goals to eliminate fossil fuel – accept it.
7) Disband the IPCC.
8) Cease government subsidies for alternative energy (wind, solar plants, biofuel, etc.).
9) Focus on mitigation strategies regardless of the source, human-caused CO_2 or natural, concentrating on regional threats.
10) Mitigation strategies include both warming and cooling scenarios – each can be deadly, but cold weather kills many more people than hot weather.
11) Allow technology research to be unimpeded by governments, except perhaps to incentivize industry.

Without interference from governments, which includes subsidizing wind, solar, bio, etc., (item 11) would eventually provide cost-effective energy alternatives based on supply and demand, and that constantly reliable factor – human innovation. If alternative energy sources were developed that were as affordable as fossil fuel, in quantities that would support the growing world population, those alternatives would be adopted, without government subsidies or policies. **Governments, supported by politicians and their media acolytes obstruct the natural innovative process by demanding that the focus remain on reducing CO_2.**

History is replete with examples of scaremongering including *end of world* scenarios, but also apocalyptic predictions similar to that from AGW alarmists. At the end of the 18th century, Thomas Malthus described such a scenario in which an exponentially growing population would outstrip food supply which was increasing only linearly. Figure 7-2(a) shows Malthus' prediction; *famine* followed the intersection of the curves. Clearly his prognosis was based on a static model that didn't consider human innovation such as advances in agricultural and scientific processes (agronomy and engineering) that substantially increased productivity. At the time of his prediction the world population was about 3 billion; today it is more than 7 billion. Figure 7-2(b) replaces "Population Growth" with "Global Temperature," and "Food Supply" with "CO_2." In this scenario, Armageddon follows the intercept of the curves – AGW alarmists call this the *tipping point*! The same curve, the same alarmism – have we learned nothing?

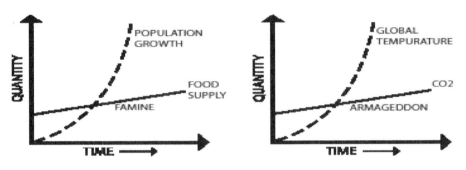

Figure 7-2 (a) Population Growth vs Food Supply; (b) Temperature vs CO_2

In his book, Al Gore says that future generations will look back and say, *"what were our parents thinking*?" I agree, but it will be targeted at the lemming-type behavior of humankind that, gripped in groupthink and tribalism, missed an opportunity to mitigate the impact of inevitable evolutionary climate change because it allowed itself to be manipulated by ideologues.

Chapter 8
Conclusions

In physics, though Einstein failed to develop a *Unified Theory of Everything* it will, one day, be accomplished; a *Unified Theory of Climate* is doubtful.

This section begins with a reminder of IPCC and U.S. government quotes that, regardless of the *science is settled* nonsense, demonstrates great uncertainty in key areas of climate analyses and modeling.

Clouds and Aerosols:

"The single greatest source of uncertainty in the estimates of climate sensitivity continues to be clouds..."

And amid this *"single greatest source of uncertainty"* admission by IPCC scientists is the statement by Dr. John Holdren, President Obama's science advisor for eight years," *...a mere one percent increase in cloud cover would decrease the surface temperature by 0.8°C."* According to NASA, clouds cover about 67% of the Earth at all times, so based on Holdren's statement **an imperceptible variation in cloud cover could account for the entire 20th and 21st century global warming.**

Water Vapor:

"Our study confirms the existence of a positive water vapor feedback ... may be weaker than we expected."

Climate models and CO_2 sensitivity claims depend on water vapor amplifying CO_2 effectiveness by a factor of three. Without such a multiplier or even a *weaker* one there is no AGW case.

Carbon Cycle:

"One of the largest unknowns in understanding the greenhouse effects is the role of oceans as a carbon sink."

And oceans cover more than 70% of the Earth's surface. Then there's the *missing sink*; measurements show that only about three-quarters of the CO_2 being produced, naturally and human-caused, is accumulating in the atmosphere and oceans. The remaining 25% cannot be accounted for; no one *knows* where it is going.

Models:

"Cloud modeling is a particularly challenging scientific problem.

The challenges remain daunting... models remain primitive....

Many cloud processes are unrealistic in current GCMs....

Continuing weakness in these parameterizations affects not only modeled climate sensitivity, but also the fidelity with which these other variables can be simulated or projected."

The above simply says the models, on which the entire AGW case is based, don't work.

Dr. Dyson said it better: *"They* **(the models)** *... do not begin to describe the real world.... They are full of fudge factors."*

It is generally accepted that over the past hundred plus years the average global temperature increased by about 0.8°C and that over that period CO_2 increased by about 120 ppm (from 280 to 400 ppm). Of the 120 ppm increase in CO_2, human contribution was about 5%; i.e., fossil-fuel burning, and other human activities added about 6 ppm. From this the IPCC and its AGW advocates claim *with certitude* a cause-and-effect relationship between the 6-ppm human contribution to CO_2 and the increase in global temperature; i.e., their case boils down to the difference between 400 ppm and 394 ppm!

Throughout the book I have dignified the AGW position by referring to it as a *hypothesis*. It is however, no more than an unsupportable *claim*. A hypothesis is testable; their position is not. The claim, shown in chapters 2, 3, 5 and 7 to be invalid, has however, among supporters, become a *belief*. If it was scientifically supportable (testable, repeatable, observable) there would no need for a *belief* and there would be no need for the propaganda and manipulation described in Chapter 1 or the deceit, lies, corruption, and abandonment of scientific integrity revealed in Chapter 2.

There is no *climate system* scientist *per se*; as discussed in Chapter 1, there are dozens of scientific disciplines that independently conduct research in one or two areas, generally without regard for other disciplines. Results are collated, *statisculated*, *corrected*, and entered into climate models by physicists or programmers. **Statisticians and modelers are the adjudicators of climate model results and forecasts, not climate scientists.**

Chapter 3 showed that while CO_2 increased linearly over the 20th century, temperatures fluctuated from warnings about an impending ice age in the 1970s to warnings in the 1980s and 1990s that global warming was reaching uninhabitable conditions, to a 15-year *pause* from 1998 to 2014 when the average global temperature stabilized; i.e., there were numerous times when the claim was not supported by observations. Albert Einstein said, *"No amount of experimentation can ever prove me right; a single experiment can prove me wrong."*

IPCC models fail to replicate historical relationships between CO_2 and temperature. Chapter 5 shows that model output disagrees with observations. And as Richard Feynman says, **"If it disagrees with observations – it's wrong. That's all there is to it."** The models are invalid, but they are the basis for the western world's oppressive energy policies and the primary tool used by alarmists to create hysteria among the public.

Climate change aka global warming is not a scientific debate; rather it is a political ideology, a *cause célèbre* foisted onto a non-critical thinking public, advocated by factions who seek a new world order, one that is governed by a central, unelected body whose goal is a form of totalitarianism.

AGW ideologues use political and media manipulation to frame the minds of an uninformed public, employing propaganda techniques intended to first denigrate those who disagree, they are *deniers*, then create fear, *its worse that ISIS*, and then provide a path to safety, *abolish fossil fuel*; it just requires sacrificing liberty and lifestyle.

Since its inception in 1988, the IPCC has demonstrated its ineptness; from the Hockey Stick scam, Climategate, nepotistic lead-author assignments and peer reviews, altering Working Group scientific results, etc. Its *raison d'être* is not to determine the cause of global warming for it has in its charter the *a priori* conclusion that anthropogenic CO_2 *is* causing global warming. It has at its foundation, an unverifiable *claim*, a key criterion necessary to create pseudo-secular religious tribalism among followers. It should have no credibility, but political and media zealots employing Alinsky-style messaging tactics and Orwellian mind-control processes galvanize an unwitting public, and silences opposition.

There are those who advocate the "Precautionary Principle," basically asking *"why not reduce CO_2, what harm is there?"* Well, there are many problems with implementing the Precautionary Principle when the science is so clearly in question: the cost in both dollars and quality of life for millions of people who need energy in quantities that cannot, at this time, be provided by non-fossil fuel. The immorality of potentially squandering trillions of dollars while known diseases, illnesses, poverty, etc., killing millions a year, are ignored. And there is no clearer admittance that the IPCC et al have no real sense that *the science is settled* than their employment of **the Precautionary Principle** which **is only used in the** *absence* **of scientific certainty**.

Given that the majority of the world's nations will not participate in fossil fuel reduction policies, and population growth will increase demand, there is no reasonable conclusion other than fossil fuel use will continue to increase over the next several decades. And, with the vast shale oil and natural gas reserves discovered worldwide, it is likely that fossil fuel will be the primary energy source for the next hundred years or more; there is just no practical alternative in the foreseeable future.

Furthermore, from a climate perspective there is no need to: first, the climate over the past century is neither unique nor abnormal in the Earth's evolution; second, the evidence shows that **reducing or even eliminating human-caused CO_2 will reduce average global temperatures by less than 0.1°C**, and; third; about 70% of CO_2 impact has already occurred – adding more, has a rapidly diminishing effect on global temperature. We have no crisis, only hysteria! Hysteria created and fostered by those who seek to control the global economy.

Final thought: While my *take-away* objectives were (1) to create uncertainty in the minds of AGW activists about CO_2 being the cause of global warming and climate change, and (2) to convince those skeptics who believe the entire debate is a hoax, that global warming is in fact occurring, I am under no illusion that such an outcome will happen.

In addition to the discomfort of uncertainty mentioned in the Preface, a major problem with expecting people to review new information and arrive at a reasoned, coherent position is in part that we are subject to psychological biases, two of which, the confirmation and recency biases, were discussed in Chapter 1 and numerous others throughout this book. Another, and one of the more egregious and least understood biases, is often called the *backfire* bias, the term first coined by Brendan Nyhan and Jason Reifler. This behavior is generally associated with ideological beliefs which, for many, include anthropogenic global warming and climate change. The backfire effect is a name for the finding that, given evidence against their beliefs, people can not only reject the evidence but also believe even more strongly in their original position. So, an irony is that while my take-away goal following reading this book was a hope for less certainty, I have probably had the opposite effect and *increased* the level of certitude held by supporters (*tribal members*) of the AGW hypothesis and those who believe the entire debate is a hoax.

However, while I'm convinced that human generated CO_2 is not the primary cause of global warming or climate change, *my* uncertainty as to the cause(s) of global warming remains, but I'm comforted by the words of Dr. Richard Feynman, one of America's greatest scientists, who said, *"The alternative to uncertainty is authority against which science has fought for centuries."* And I offer the following aphorism:

Only fools or tyrants assert with dogmatic infallibility that they *know* the cause(s) of global warming.

For all inquiries or comments contact Mike Sangster at:
michaelsangster@comcast.net

Printed in Great Britain
by Amazon

48427250R00165